HIGH-LEVEL DOLLARS, LOW-LEVEL SENSE

A Critique of Present Policy for the Management of Long-Lived Radioactive Wastes and Discussion of an Alternative Approach

HIGH-LEVEL DOLLARS, LOW-LEVEL SENSE

A Critique of Present Policy for the Management of Long-Lived Radioactive Wastes and Discussion of an Alternative Approach

by

Arjun Makhijani
Scott Saleska

A Report of the Institute for Energy and Environmental Research.

The Apex Press is an imprint of the Council on International and Public Affairs, 777 United Nations Plaza, New York,
New York 10017 (212/953-6920).

Library of Congress Cataloging-in-Publication Data

Makhijani, Arjun.
    High-level dollars, low-level sense : a critique of present policy
for the management of long-lived radioactive wastes and discussion
of an alternative approach / by Arjun Makhijani and Scott Saleska.
        p. cm.
    "A Report of the Institute for Energy and Environmental Research."
    ISBN 0-945257-42-2 : $12.95
        1. Radioactive waste disposal—Government policy—United States.
    2. Radioactive waste disposal—Government policy—Sweden.
    3. Radioactive waste disposal—Environmental aspects—United States.
    4. Radioactive waste disposal—Environmental aspects—Sweden.
    I. Saleska, Scott.   II. Institute for Energy and Environmental
    Research (Takoma Park, Md.)   III. Title.
    TD898.14.G68M35 1992
    363.72'89056—dc20   91-40406
                                                                      CIP

Cover design by Janette Aiello
Typeset and printed in the United States of America

CONTENTS

PREFACE

The management of nuclear waste, which contains materials that remain hazardous for up to millions of years, is one of the most vexing, contentious, and costly environmental issues of our time. Nuclear waste management has been plagued with failures, poor science, and unanticipated environmental events—such as rapid migration of radioactive contaminants from the soil into groundwater—which have made a mockery of many a computer model.

The scene is also littered with institutional and regulatory failures and absurdities. In the United States today, nuclear wastes are classified, not so much according to the threat they pose to human health or the environment, but according to the process which produced the waste. For example, a catchall category called "low-level" waste contains some components which are more radioactive than some "high-level" waste. Some wastes have health criteria which govern their disposal. Other wastes do not. Some wastes are designated as suitable for shallow-land burial. Other wastes of comparable danger are designated for disposal in a deep underground repository. There are no adequate programs to address whole categories of other wastes of comparable danger, notably soil contaminated with plutonium and other long-lived radioactive elements, which may, by default, be left lying around endangering public health and the environment for thousands of years.

This welter of problems, along with concern for future generations and environmental degradation in general, has given rise to vigorous public opposition to nuclear waste disposal sites wherever they have been proposed in recent years. Such opposition does not derive simply from a "not in my backyard" syndrome. That syndrome does play a role and it is understandable in view of the long-lived nature of the threat. People's fears, however, also stem from the problems which have arisen from the basic conflicts of interest in the institutions—notably the Department of Energy and the Nuclear Regulatory Commission—which have regulated nuclear waste disposal. These institutions have an agenda—sometimes stated, and sometimes only implicit—of producing nuclear weapons and promoting nuclear power. For more than a decade, the executive branch of the U.S. government has also explicitly and vigorously pursued that same agenda, coloring the actions of the institutions which operate under it. Under present institutional arrangements, these goals have been in basic conflict with providing sufficient time and resources to protect future generations as best we can from a considerable threat to the environment which our activities have created in the form of long-lived nuclear waste.

From these conflicts of interest have arisen failures which have been costly both to the environment and to the public purse. We undertook this work in order to discuss the failures in all areas of nuclear waste disposal, focusing especially on the problem of classifying nuclear wastes in a manner that corresponds to the threats that they pose. This has enabled us to propose a unified approach to the management of the problems that cuts across current waste categories. We also can see clearly the need to minimize generation of long-lived radioactive wastes.

Proponents of quick land-based disposal of nuclear waste often resort to scare tactics in order to push new disposal sites on the public. These range from a purported need for more nuclear power plants to threats that huge portions of the medical care system may shut down if there are no new disposal sites. Such tactics create a perceived urgency which does not arise from any technical problem. There are ways to provide for interim storage of nuclear wastes which pose far smaller threats than quick land-based disposal and hurried transportation. There are also ways to minimize use of long-lived radioisotopes, especially in medicine, and the medical community has begun taking steps over the last many years in that direction. The perceived urgency arises more from artificial deadlines that Congress and governmental agencies have created, largely in response to pressure from industry to quickly dispose of the wastes. Such artificial deadlines can and should be changed, so that the environment, public health and the public purse may be better protected.

A few years ago, parts of the chemical industry were given to painting similar scare scenarios about chlorofluorocarbons, which are destroying the earth's protective layer of stratospheric ozone. It was stated that we may have to give up refrigerators and computers and automobile air conditioners if CFCs were reduced by even 50 percent by the year 1998. The threat to the ozone layer from these compounds has proved to be so severe, however, that it has clearly become necessary to phase out these chemicals altogether, and other ozone-depleting chemicals besides. Today, under pressure from an international treaty which requires the total phase-out of these chemicals by the year 2000, as well as more stringent local laws, many industries now find that they can get rid of them well before 1995! Moreover, many of them are saving money as they do it.

Careful consideration of alternative energy sources, energy conservation, the use of short-lived radionuclides in medicine and research, and the end of the Cold War may enable a phase-out of the generation of long-lived radioactive wastes except for some minor medical and research uses. While such considerations are beyond the scope of this book, we urge that they be taken up as a matter of high priority in public policy, even as we hope to put the attempts to address the problems of the long-lived radioactive wastes which already exist on a sounder technical and institutional footing. The latter is the subject of this enquiry. These issues must be addressed both out of concern for the protection of public health and the environment and as a matter of financial prudence.

This study was funded in part by the Nuclear Waste Projects Office of the State of Nevada, which operates in part under Department of Energy contract DE-FG-08-85-NV10461. The rest of the funds for its preparation came from general support funds provided to the Institute for Energy and Environmental Research (IEER) by the Public Welfare Foundation and from IEER's own institutional funds and from a gift of Dr. Gopi B. Makhijani.

We would like to thank Steve Frishman of the Nevada Nuclear Waste Projects Office; Don Hancock of the Southwest Research and Information Center; Charles Hollister, of the Woods Hole Oceanographic Institution; Kemp Houk of Don't Waste U.S.; Judy Treichel, of the Nevada Nuclear Waste Task Force; and Kitty Tucker of the Health and Energy Institute for reviewing this study. Their many and insightful comments enabled us to publish a greatly improved final product. We are especially indebted to Don Hancock for his suggestion that we be more comprehensive in our approach to the problem of radioactive waste, and for the many detailed and constructive suggestions that he made towards that end.

During the research phase of this project we were greatly and ably assisted by the efforts of former IEER staff member Deborah Landau, to whom we owe a debt of thanks. We are also greatly indebted to David Dembo of The Apex Press, who worked with us during the final stages to convert the manuscript into a published book.

Those acknowledged here do not necessarily in any way endorse the findings, conclusions, or recommendations of this study, the responsibility for which lies solely with us. We also, of course, take full responsibility for any errors.

Arjun Makhijani

Scott Saleska

Takoma Park, Maryland

October 1991

Chapter 1

INTRODUCTION

The management and disposal of long-lived radioactive wastes —the great majority of which are the result of nuclear weapons production and commercial nuclear power generation—has been a technical and political problem for many decades.

There are many types of radioactive wastes, varying in radioactivity level, longevity, and hazard. Much (although by no means all) radioactive waste is subsumed under two broad categories, named "high-level waste," and "low-level waste." The other principal categories of nuclear waste are transuranic waste, and uranium mill tailings. Although many of the attempts to address the "nuclear waste problem" have focused on one or another of the above categories, it is a principal thesis of this study that these categories are fundamentally misconceived, and that this misconception has led to many of the problems that continue to exist for nuclear waste disposal.

One common factor for all categories of nuclear waste is the presence of at least some amount of long-lived radionuclides. It is on the management and disposal of these long-lived components that this study focuses, regardless of which official waste "category" such components happen to fall into.

In addition to examining the characteristics of some of the most hazardous and long-lived waste forms, we have also addressed the question of the adequacy of current policies for managing them. We identify fundamental problem areas in the technical, regulatory, and managerial aspects of present programs, and suggest an alternative structure to correct deficiencies in each of these areas.

To this end, this book is organized in the following manner. Chapter 2 provides an overview of the radioactive waste problem, including the origin of nuclear wastes, and the characteristics of each of the currently defined radioactive waste categories, along with the amounts and locations of the waste. Chapter 3 contains an explanation and critical analysis of the various components of the current approach to management of these wastes. Chapter 4 lays out our proposal for an alternative, integrated approach to waste management that addresses many of the deficiencies and shortcomings which we identify in Chapter 3. Finally, our findings and recommendations are summarized in Chapter 5.

Chapter 2

OVERVIEW OF THE RADIOACTIVE
WASTE PROBLEM

The current policies for the management and disposal of radioactive waste in the U.S. are the subject of considerable controversy and disagreement. This is true for all of the principal official categories of commercial radioactive waste in the U.S., i.e. for "high-level," "low-level," and "transuranic" radioactive wastes, as well as for uranium mill tailings. However, before discussing the details of these policies and controversies—and our proposed resolution of them—this first chapter will provide an overview of the origins of radioactive wastes and some of the problems posed by them.

The first section of this chapter briefly describes the nuclear fuel cycle and how it generates radioactive waste. The second section reviews in greater detail the sources and characteristics of each of the official categories of radioactive waste.

A. Radioactive Waste and the Nuclear Fuel Cycle

Radioactive waste is produced by a number of sources, but by far the largest quantities of it—in terms of both radioactivity and volume—are generated by the commercial nuclear power and military nuclear weapons production industries, and by nuclear fuel cycle activities to support these industries such as uranium mining and processing.

Commercial Power Generation

The set of activities which begins with uranium mining and ends with spent fuel waste is called the nuclear fuel cycle. As Figure 1 shows, the nuclear fuel cycle begins with uranium mining. The wastes from mining and initial processing contain some uranium and most of its radioactive decay products, notably thorium-230, and radium-226 and its decay product radon (a gas). These wastes are known as "mill-tailings."

Before being used in a reactor as fuel, the uranium is typically taken through several additional processing stages. Central to this preliminary processing is uranium enrichment, which makes the uranium more usable in current power generators. This is accomplished by increasing the concentration of the fissionable isotope of uranium, uranium-235, relative to the non-fissionable isotope, uranium-238.[1]

When uranium comes from the milling process, it is in the form of uranium oxide (U_3O_8), which is often called "yellowcake." Currently employed enrichment technologies require that the uranium be in the chemical form of UF_6, or uranium hexafluoride. Thus, before it can be enriched, uranium must go through the process called "conversion," which in the U.S. is done at one of two conversion plants.[2] From the conversion plant the uranium is shipped to an enrichment plant.

After enrichment, the uranium undergoes a fuel fabrication process in which it is formed into pellets and put into long fuel rods. Bundles (called "assemblies") of these fuel rods are loaded into nuclear power reactors.

When the enriched uranium fuel is irradiated with neutrons in nuclear reactors, a sustained nuclear chain reaction takes place. This nuclear reaction consists primarily of the fission (or splitting apart) of uranium-235 atoms to yield energy in the form of heat and radioactive

---

1. Natural uranium is almost 99.3 percent U-238, 0.7 percent U-235, and a tiny fraction U-234. U-234, though a small fraction by weight, contributes as much radioactivity to ore as U-238, due to U-234's much shorter half-life. Typical U.S. nuclear power reactors use uranium fuel that has been enriched to the point where it contains several percent (3 to 4 percent) uranium-235. The uranium for the reactors which produce plutonium for nuclear weapons has a different mix of U-238 and U-235 than commercial power reactors.

2. The two are the Sequoyah Fuels Corporation's plant near Gore, Oklahoma (formerly owned by Kerr-McGee Corporation, but recently bought by General Atomics); and the Allied-Chemical conversion plant near Metropolis, Illinois.

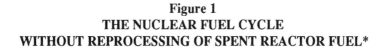

Figure 1
THE NUCLEAR FUEL CYCLE
WITHOUT REPROCESSING OF SPENT REACTOR FUEL*

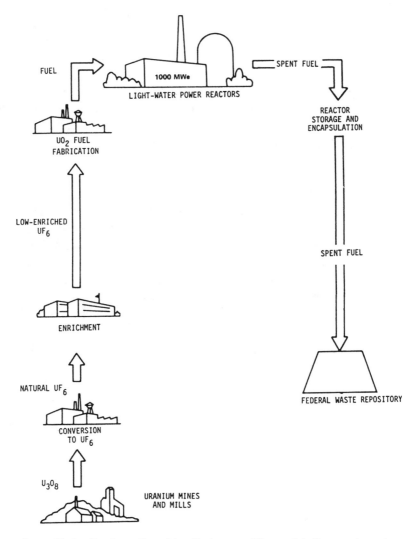

Source: Nuclear Regulatory Commision, *Environmental Survey of the Reprocessing and Waste Management Portions of the LWR Fuel Cycle*, W.P. Bishop and F.J. Miraglia, eds., NUREG-0016, Washington, D.C.: NRC Office of Nuclear Material Safety and Safeguards, October 1976, as cited in Lipshutz 1980, p. 30.

* UF_6=Uranium hexaflouride; U_3O_8=Uranium oxide ore; UO_2=Uranium dioxide; MWe=Megawatts electric power

fragments called "fission products." (Cesium-137 and strontium-90 are prominent examples of fission products in irradiated nuclear fuel.) As the reaction proceeds, more and more of the uranium-235 in the fuel rods is gradually transformed into these fission products.

Some of the neutrons inside the reactor, instead of causing more uranium-235 atoms to split, are absorbed by other atoms, causing them to be converted into a different type of atom. In this way, parts of the fuel assemblies, components of the reactor, or even materials suspended in the reactor cooling water are made radioactive. Materials made radioactive in this way by neutron absorption are called "activation products" (typical examples of activation products are cobalt-60 and carbon-14). This process also results in the production of plutonium-239 (the isotope of plutonium which is used to make nuclear weapons) when uranium-238 atoms in the fuel rods absorb neutrons.

After several cycles of "burn-up" time, the irradiated fuel rods, known as "spent fuel," are removed from the reactor core.[3] A typical cycle may be 12, 18, or 24 months, and a given batch of fuel rods may remain in the reactor core for two or three cycles. Because of the build-up of the radioactive fission and activation products, the spent fuel assemblies are millions of times more radioactive than before they were placed in the reactor. They are also very hot thermally. This spent fuel is therefore stored in pools of cooling water at reactor sites. There are currently slightly over 110 licensed nuclear power reactors in the U.S., located at about 70 nuclear plant sites.

It should be noted that, while it is no longer done in the U.S. commercial power sector because of the economic and environmental difficulties, spent fuel from power reactors can be processed chemically to recover the unused uranium and newly created plutonium for use in new fuel elements. This is known as reprocessing.

---

3. "Burn-up" is a technical term used to express the amount of energy extracted from a given mass of nuclear fuel. Burn-up is generally measured in units of megawatt-days of thermal energy extracted per metric ton of uranium (MWd/MTU). Burn-up potential is roughly correlated with the enrichment level of the fuel. Average burn-up levels in the U.S. in 1989 were 27,165 MWd/MTU for boiling water reactors, and 35,255 MWd/MTU for pressurized water reactors. (DOE 1990d, p. 13)

Nuclear Weapons Production

Nuclear weapons production consists of a complex series of steps that produce nuclear material, fabricate this material into components for nuclear weapons, assemble the components, and test the manufactured weapons. The responsibility for U.S. nuclear weapons production rests with the Department of Energy (DOE), which runs the numerous facilities at over a dozen major sites which make up the U.S. weapons production complex.[4]

The operations of the U.S. nuclear weapons production complex have been referred to by at least one federal office as "potentially one of the more dangerous industrial operations in the world."[5] The activities of the weapons complex result not only in the generation of vast quantities of radioactive waste, but also in a wide variety of chemically hazardous wastes as well.

At the center of modern nuclear weapons production activities has been the creation of the materials plutonium-239 and tritium, a radioactive form of hydrogen. These materials are created by neutron bombardment, and since operating nuclear reactors produce large amounts of neutrons, the U.S. has typically used nuclear reactors to produce these nuclear materials.

When used for commercial power generation, a nuclear reactor's desired product is heat, because it can be used to drive a turbine and generate electricity. As discussed in the section on commercial power, above, the generation of heat in a nuclear reactor also results in the buildup of radioactive by-products which are treated as waste. In the production of nuclear materials for weapons, however, the opposite is the case: some of the radioactive materials produced (plutonium and tritium in particular) are the sought-for product, while the heat generated is extraneous by-product.[6]

---

4. For more technical details on nuclear weapons components and how they are produced, see NRDC 1987.
5. GAO 1986b, p. 8.
6. Note, however, that some reactors (the N-reactor at the Hanford nuclear reservation, for example) have served as dual-purpose reactors, both producing nuclear materials for weapons and generating electricity at the same time. One of the designs (the high-temperature gas-cooled reactor) that the DOE is currently considering for a new production reactor for the weapons materials tritium and plutonium, is a dual-purpose reactor as well.

While reprocessing is not essential to nuclear power, it is essential to producing plutonium for nuclear weapons. After irradiation, the fuel rods contain uranium, fission products, various isotopes of plutonium, and other "transuranic" elements (i.e., elements with atomic numbers higher than uranium—which has atomic number 92—in the periodic table of elements).

Reprocessing is necessary to recover the plutonium in the fuel rods so that it can be used for making nuclear weapons. During reprocessing, the spent fuel rods are dissolved in an acid bath so that the plutonium and uranium can be removed. What remains is a highly radioactive and very thermally hot liquid waste stream which still contains essentially all of the fission and activation products.[7] This liquid stream is classified as "high-level" radioactive waste by federal regulations.

Reprocessing and subsequent handling and machining of the extracted plutonium and other transuranics by military nuclear plants also results in the contamination of a wide variety of waste materials with transuranic nuclides. These are classified as transuranic wastes.

At every stage of the nuclear fuel cycle in both the military and commercial power sectors, a diverse array of waste materials contaminated with varying levels of radioactivity are generated. Any of these wastes that do not fall into any of the radioactive waste categories described above are lumped into a catch-all category and designated as "low-level" wastes.

B. Radioactive Waste Characteristics

This section will discuss in greater detail the characteristics, locations, and amounts of each of the currently defined categories of radioactive waste. As discussed in the section above, these categories are:

- high-level wastes, including both commercial spent fuel and military reprocessing wastes;

---

7. Note, however, that some radioisotopes, such as krypton-85, carbon-14, iodine 131, and a few others are typically in gaseous form and do not remain in the liquid waste stream after reprocessing. They can be trapped, but in the past, they were often vented to the atmosphere in large quantities.

- transuranic wastes, which today come primarily from the Department of Energy's military production activities;

- uranium mill tailings from the processing of uranium ore; and

- low-level wastes, which are generated by both the commercial and military sectors.

Wastes from decommissioning and shutting down old nuclear reactors, because they pose special problems and represent such a significant amount of the future radioactive waste generation, are discussed in a separate section, even though they are also technically classed as "low-level" wastes. We also include a brief separate section on "other wastes" which warrant mention. These include "mixed wastes," an overlapping category which includes radioactive wastes that are mixed with chemically hazardous components. Dealing with such wastes is complicated by the fact that they are subject to two sets of sometimes conflicting regulations: radioactive waste regulations and hazardous waste regulations.

Before proceeding we should note that the amounts of waste generated by the DOE's military production activities are typically subject to greater uncertainties than the amounts from the commercial sector. This is because, in many cases, complete records are lacking for wastes in the DOE complex, and DOE officials do not know how much waste has been disposed of as a result of past operations—or even how many military waste dump sites exist.[8] Further, due to environmental and safety problems which plague the weapons complex, the politically and technically controversial aspects of many DOE plans, and the uncertain military demand for continued U.S. weapons production in light of the apparent end of the cold war and reductions in nuclear weapons, it is unlikely that the weapons complex in the future will continue in the same fashion as it has in the past. These complications will unavoidably affect the plans for generation, management, and disposal of radioactive wastes throughout the nuclear weapons complex.

------

8. GAO 1988c, pp. 3, 9.

High-Level Wastes: Spent Fuel and Reprocessing Wastes

Spent Fuel

When the chain reaction in a commercial reactor is stopped, the fuel remains intensely hot. This is because heat continues to be generated by the radioactive decay of fission products. These fission products have half-lives that range from a fraction of a second to millions of years. The ones that decay most rapidly generate the most heat at first, being the principal source of heat in cases of melt-down accidents, such as the one that occurred at Three Mile Island in 1979.

After removal from the reactor, the fuel is so hot that it must be stored underwater in spent fuel pools for a considerable period, with the water circulating constantly. The characteristics of spent fuel in spent fuel pools from pressurized water reactors (PWRs, the most common type) and boiling water reactors (BWRs) are similar, but not exactly the same. Figures 2 and 3 show, as a function of time, the radioactivity in spent fuel which initially contained one metric ton of uranium (abbreviated as MTU) for BWRs and PWRs.

One year after withdrawal from a pressurized water reactor, the radioactivity in a metric ton of uranium irradiated for 30,000 megawatt-days is almost 2 million curies. In 10 years, this is down to about 300,000 curies. In 100 years, it is about 30,000 curies. The heat generation approximately follows the radioactivity. It goes down from about 7,000 watts per metric ton of uranium after one year to about 1,000 watts after 10 years and 200 watts after 100 years. The figures for a boiling water reactor are similar, as shown in Figures 2 and 3.

Different radionuclides are important at different times. The short half-life fission products, like iodine-131 (half-life, eight days), dominate the health threats early on. Other elements, like ruthenium-106 become relatively more important at intermediate times (on the order of one year). For longer time frames, three kinds of radioactive isotopes are important:

1. Krypton-85 (half-life, 10.7 years), cesium-137 (half-life 30.2 years) strontium-90 (half-life 28.8 years) and plutonium-241 (half-life 14.4 years). These elements constitute the bulk of the radioactivity a few to a few hundred years after discharge from the reactor. (Plutonium-241 decays into other radioactive elements, called "daughter products," with much longer half-lives).

2. Very long-lived beta and gamma radiation emitting elements, in-
    cluding carbon-14 and long-lived fission products like technetium-
    99, iodine-129 and cesium-135 which have half-lives of thousands
    of years to millions of years;

3. Long-lived alpha radiation emitting elements like radium as well as
    transuranics like plutonium-239.

Krypton-85, strontium-90, cesium-137, and plutonium-241 present
special threats because, after a few years, they represent the largest
amount of radioactivity compared to all the other radionuclides. Further, strontium and cesium mimic calcium and potassium, respectively,
in the human body and can replace them. For instance, radioactive
strontium accumulates in the bone, increasing the risk of bone cancer
and leukemia.

One important distinction between isotopes of the same element
with long and short half-lives is that the radioactivity per unit weight is
approximately inversely proportional to the half-life. For example,
about 76,000 times more cesium-135 (half-life 2.3 million years) is
required to produce the same amount of radioactivity (and thus, potential health damage) as a given quantity of cesium-137 (half-life 30.2
years).

The projected total commercial spent fuel accumulation through
1991 is about 24,000 metric tons.[9] Virtually all of this is stored on site
near the reactor from which it came. To allow for planning for future
spent fuel management and disposal, the DOE has made projections of
the amount of spent fuel expected in the future. Total projected accumulation of commercial spent fuel under the DOE's basic "no-new-orders" scenario is about 86,800 metric tons.[10] DOE also considers an
"upper reference" case in which it is assumed that commercial nuclear
power generating capacity will continue to grow. In planning for spent

---

9. Harvey 1991.
10. This scenario anticipates that no new reactors will be ordered, but that reactors
    currently in operation or under construction will operate to the end of their licensed
    40-year lifetimes. This is called the "no-new-orders end-of-reactor-life" scenario.
    According to this scenario, the last spent fuel is projected to be discharged in the
    year 2037. (DOE 1990a, p. 8, citing DOE Energy Information Administration
    forecasts.)

Figure 2
BOILING-WATER REACTOR SPENT FUEL*

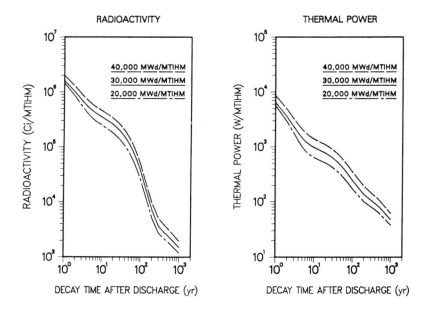

Source: DOE 1990d, p. 21.
   * Radioactivity and thermal power of 1 metric ton of heavy metal of BWR spent fuel as a
function of burn-up and time from reactor discharge. MTIHM=Metric Tons Initial Heavy Metal. Equivalent to Metric Tons Uranium (MTU). A measure of the amount of nuclear fuel.
Ci/MTIHM=Curies per metric ton of fuel. A measure of the radioactivity in spent fuel.
W/MTIHM=Watts of heat per metric ton of spent fuel. A measure of the amount of thermal
power (heat) given off by spent fuel. MWd/MTIHM=Megawatt-days per metric ton of spent
fuel. A measure of fuel "burn-up," or the amount of energy extracted from a given amount of
nuclear fuel. The greater the burn-up, the greater the radioactivity and heat of the fuel.

Figure 3
PRESSURIZED-WATER REACTOR SPENT FUEL*

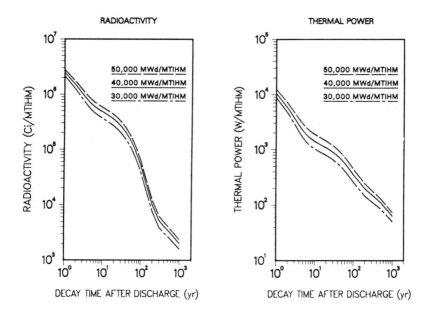

Source: DOE 1990d, p. 21.

* Radioactivity and thermal power of 1 metric ton of heavy metal of PWR spent fuel as a function of burn-up and time from reactor discharge. MTIHM=Metric Tons Initial Heavy Metal. Equivalent to Metric Tons Uranium (MTU). A measure of the amount of nuclear fuel. Ci/MTIHM=Curies per metric ton of fuel. A measure of the radioactivity in spent fuel. W/MTIHM=Watts of heat per metric ton of spent fuel. A measure of the amount of thermal power (heat) given off by spent fuel. MWd/MTIHM=Megawatt-days per metric ton of spent fuel. A measure of fuel "burn-up," or the amount of energy extracted from a given amount of nuclear fuel. The greater the burn-up, the greater the radioactivity and heat of the fuel.

fuel disposal under the upper reference scenario, DOE uses a cutoff date of 2020, at which point about 96,900 metric tons of spent fuel are assumed to have been generated.[11] Military high-level wastes from reprocessing DOE's spent fuel are in addition to these quantities, and are discussed below.

Reprocessing Waste

High-level waste from reprocessing operations is present at four locations in the United States: the Savannah River Site in South Carolina; the Hanford Nuclear Reservation in Washington State; the Idaho National Engineering Laboratory; and at West Valley, New York.

At the first two locations, the wastes are mainly from plutonium production for nuclear weapons. At Idaho, they are wastes from reprocessing of spent fuel from naval reactors; at West Valley they are the wastes from the reprocessing of some commercial spent fuel and some military spent fuel that was done there between 1966 and 1972. The reprocessing plant at West Valley has been shut down since 1972.

Figure 4 shows the volume and radioactivity of high-level wastes at the various sites through 1989. As the figure shows, Hanford has a greater volume, but about 60 percent of the total radioactivity is in the wastes at Savannah River.

The wastes are stored in various forms at these sites. At Savannah River the wastes are in the form of liquids, sludge, and salts resulting from evaporation. At Hanford, in addition to these waste forms, there are "capsules" of separated cesium-137 and strontium-90. At Idaho there are some high-level liquids and powder resulting from calcining liquid high-level waste. At West Valley, the waste consists of liquids and sludge.

The DOE, which is responsible for managing all of these reprocessing wastes, plans eventually to process part of them into molten glass and cast the mixture into large cylinders which will be disposed of along with commercial spent fuel. (Significant fractions of the wastes, however, are expected to be separated out before glassification and classed as low-level wastes for shallow-land burial.) This process is known as glassification or vitrification. The glass used is borosilicate glass,

---

11. *Ibid.*

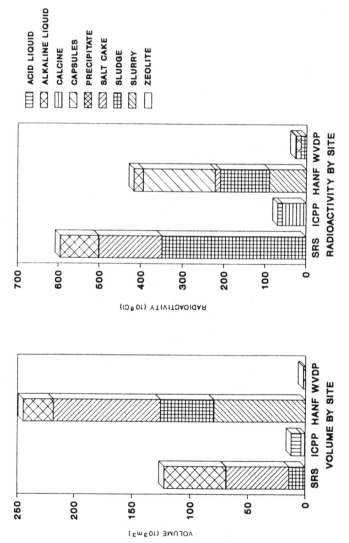

Figure 4

DISTRIBUTION OF TOTAL ACCUMULATIVE VOLUME AND RADIOACTIVITY OF HLW BY SITE* AND TYPE THROUGH 1989

Source: DOE 1990d, p. 41

* SRS=Savannah River Site (Aiken, South Carolina); ICPP=Idaho Chemical Processing Plant (Idaho Falls, Idaho); HANF=Hanford Nuclear Reservation (Richland, Washington); WVDP=West Valley Demonstration Project (West Valley, New York).

which is similar to the Pyrex glass used to make kitchenware. A vitrification plant—called the Defense Waste Processing Facility—has been built at Savannah River, and although the DOE originally planned to begin converting reprocessing wastes to radioactive glass in 1990,[12] problems have caused delays. The DOE currently expects to open the Savannah River vitrification plant in 1993, although further delays may be possible. A vitrification plant is under construction at West Valley, and the DOE plans to begin full-scale glassification there in 1996.[13]

There have been plans to do the same at Hanford. However, despite the DOE's nominal start-up date of 1999, diverse problems have put the actual Hanford vitrification plant and schedule in limbo. These problems include questions about the safety and adequacy of the plants which would process the wastes prior to vitrification, potential operating problems for the vitrification plant due to the presence of potentially explosive chemicals in the waste, and, as the DOE puts it, "lack of detailed knowledge of tank constituents" at Hanford. The DOE itself has noted that these problems "could have a significant impact on [the Hanford vitrification plant's] plans and schedule."[14]

Much of the radioactivity in high-level reprocessing waste consists of elements with half-lives on the order of one year or less. However, a considerable portion consists of cesium-137 and strontium-90, which are among the most troublesome radionuclides from the point of view of health effects and long-term management. In addition, significant quantities of very long-lived fission products and transuranics are also present.

The total quantity of high-level reprocessing wastes assumed for disposal planning purposes is 17,750 canisters of glass (from about 8,875 metric tons of irradiated fuel) due to military production activities, and 300 canisters (about 640 metric tons) from now-defunct commercial reprocessing operations at West Valley, New York.[15]

---

12. DOE 1984, p. 64.
13. DOE 1991b, p. 185.
14. *Ibid.*
15. DOE 1990a, p. 8.

Transuranic Waste

The term "transuranic" (or, "above uranium") applies to all elements whose atomic number (the number of protons in the nucleus of an atom) is greater than 92, the atomic number of uranium. The waste category defined by government regulations as transuranic waste (or "TRU waste") is somewhat more limited. "TRU waste" includes only waste material that contains transuranic elements with half-lives greater than 20 years, and which are found in concentrations greater than 100 nanocuries per gram (although as discussed below, the standard was 10 nanocuries per gram before 1984).[16] Thus, if the concentrations or the half-lives are below the limits, it is possible for waste to contain transuranic elements but not be officially designated as transuranic waste.

Although all elements up to and including uranium are found in nature, no elements with atomic numbers greater than uranium—that is, no transuranic elements—are naturally occurring.[17] Thus, transuranic elements are the artificial elements. All transuranic elements are unstable (and thus radioactive), many of them are alpha-emitters, and many (although not all) have very long half-lives.

It is because of the longevity and alpha-emitting quality of transuranics that wastes contaminated by significant quantities of these elements pose special disposal problems. For this reason, in 1970, the federal government defined a subset of what had been treated as "low-level" waste as transuranic, or TRU, wastes. Before this reclassification, these wastes had often been buried along with low-level waste in shallow burial trenches. Transuranic wastes were required to have greater confinement from the environment than "low-level" waste, although for many years this requirement was not enforced at commercial radioactive waste disposal sites.[18]

---

16. DOE 1990d, p. 75.
17. Although at least one instance is known in which a small quantity of plutonium (long since decayed away) and fission products must have been created naturally about 2 billion years ago in a "natural" reactor at an underground location in what is now Gabon, West Africa. This phenomenon was made possible by a high concentration of uranium and by the fact that the percentage of uranium-235 was much higher so long ago than the 0.7 percent found in today's uranium ores. (Eisenbud 1987, p. 171.)
18. DOE 1990d, p. 75. For background on transuranic waste at commercial disposal sites, see Carter 1987, p. 73.

The transuranic contamination level which defined transuranic waste was originally set at concentrations of radioactivity greater than 10 nanocuries per gram (nCi/gram). In 1984, the radioactivity level cut-off was redefined as 100 nCi/gram—ten times higher.[19]

Most of the transuranic waste volume consists of dirt and trash (e.g. rags, coveralls, equipment, and tools) that have been contaminated with transuranic elements during nuclear reprocessing of irradiated fuel, the machining and processing of plutonium for use in nuclear warheads, or the careless handling of transuranic materials at various points in the nuclear weapons production complex.

Since neither reprocessing of irradiated fuel nor direct machining or processing of transuranic elements is currently conducted in the commercial nuclear power sector, essentially all transuranic wastes generated in the U.S. today come from the DOE's activities in its nuclear weapons production complex. If, however, nuclear reprocessing for commercial nuclear power generation were ever to resume in the U.S., the commercial sector would also become a significant generator of transuranic waste.

Most of DOE's transuranic wastes emit primarily alpha radiation (which has very weak penetrating power), and are considered "contact-handled" transuranic wastes. For this waste, the shielding provided by the waste package itself is considered sufficient. About 2.4 percent of the transuranic waste volume also contains sufficient beta or gamma emitters to deliver radiation doses of 200 millirem per hour or more; this waste is handled more carefully and is designated "remote-handled" transuranic waste.[20] In addition, details about a small amount (less than one percent) of transuranic wastes are considered "classified information" and are not publicly available. The DOE has requested that some state environmental regulatory officials obtain security clearances to address the issue of environmental compliance for this waste.[21]

The DOE's transuranic waste as currently defined is found in primarily three forms: in retrievable storage, as buried waste, and as contaminated soil. The transuranic waste in retrievable storage is the waste which has been set aside since transuranic waste was first defined as a separate category requiring more substantial isolation. Much of

---

19. DOE 1990d, p. 75.
20. DOE 1990d, p. 75.
21. DOE 1991b, p. 186.

the transuranic waste from before this time or which was not set aside in retrievable storage was buried.

Transuranic contaminated soil has resulted from a number of factors. Leaking burial containers have resulted in transuranic contaminated soil at some sites. For example, according to one government report, about two-thirds of the drums containing transuranic waste have severely deteriorated and contaminated the surrounding soil.[22] In addition, the Atomic Energy Commission (AEC) and the DOE have dumped billions of gallons of radioactive liquid low-level and transuranic wastes directly into the soil via unlined cribs, ponds, trenches, and ditches at numerous sites around the country.[23] (This contamination can be quite significant. For example, at one point in the early 1970s, the AEC became concerned that the transuranic contamination at the bottom of one of its waste cribs at Hanford might be so great that a critical mass of plutonium might be present—posing the threat of a spontaneous chain reaction or even a low-level nuclear explosion.[24])

Table 1 lists DOE's estimates of the amounts, as of the beginning of 1990, of these three forms of transuranic waste at the sites where they are currently found. In addition to these sites, transuranic waste is generated at several other sites around the U.S., including Lawrence Livermore National Laboratory in California, and Argonne National Laboratory in Illinois. It is then shipped from these sites to those listed in Table 1. The principal generator until recently, however, has been the Rocky Flats Plant near Golden, Colorado, where plutonium metal has been machined and processed into pits for nuclear warheads. In recent years, Rocky Flats has been in a forced shut-down state, due to extensive environmental and safety problems that have come to light at the plant. Thus, transuranic waste generation there is currently substantially less than it would be if the DOE were ever able to return to full-scale production mode, as it plans to do.

As can be seen, the total amount of transuranic waste in the nuclear weapons complex, based on the DOE's current estimates, is in the range of 390,000 to 540,000 cubic meters.

---

22. GAO 1986c, p. 21.
23. GAO 1988c, pp. 9-10.
24. U.S. Atomic Energy Commission, *Contaminated Soil Removal Facility, Richland, Washington*, WASH-1520, p. 7 (April 1972), as cited in Lipshutz 1980, p. 131.

Table 1
AMOUNTS OF TRANSURANIC WASTE (IN CUBIC METERS)

Site	Retrievable Storage	Buried	Contaminated Soil	TOTAL
Hanford, WA	10,180	109,000[a]	31,960	151,140
Idaho Nat'l Engineering Laboratory, ID	37,450	57,100	56,000 - 156,000	150,550 - 250,550
Los Alamos Nat'l Lab, NM	7,420	14,000	1,140	22,560
Mound, OH			300 - 1,000	300 - 1,000
Nevada Test Site, NV	610	--	unknown[b]	> 610
Oak Ridge Nat'l Lab, TN	1,970	6,200	13,000 - 61,000[c]	21,170 - 69,170
Rocky Flats Plant, CO	790	--	unknown	> 790
Sandia Nat'l Labs, NM	--	3	--	3
Savannah River Site,SC	3,140	4,530	38,000	45,670
TOTAL VOLUME	61,560	190,840	140,400 - 289,100	393,000 - 541,500
RADIOACTIVITY (alpha curies)[d]	1,130,800	121,800	> 17,000[e]	> 1,270,000

Source: DOE 1990d, pp. 82,85.
Notes:
(a) Includes soils mixed with buried wastes.
(b) The source (DOE 1990d) does not list Nevada Test Site as containing transuranic-contaminated soil. However, some areas of the Test Site are known to contain significant plutonium contamination as a result of partial-detonation tests which scattered rather than fissioned much of the plutonium in the bombs so tested.
(c) Oak Ridge soil estimate applies if soil containing transuranic waste can be isolated from a total contaminated soil inventory of about 1.6 million cubic meters.
(d) Sum of figures reported by storage site. Does not include beta or gamma radioactivity, or radiation from decay products.
(e) 17,000 curies is average of range of alpha radioactivity reported for contaminated soil at Hanford and Mound only. Radioactivity of contaminated soil at other sites is reported as unknown.

Uranium Mill Tailings

Wastes from uranium milling are what is left over from the extraction of uranium oxide (usually in the form of U_3O_8) from uranium ore. These wastes are generally referred to as uranium mill tailings, and they often exist mixed with liquid mill effluents in a slurry or sludge form in ponds, or in dryer piles of a gray fine-grained sand-like material. These uranium mill tailings contain about 85 percent of the radioactivity of the original ore, primarily in the form of the uranium-238 decay products radium-226 (with a half-life of about 1,600 years) and thorium-230 (half-life about 80,000 years).

Although the average concentration of radioactivity in mill tailings is fairly low in comparison to other categories of nuclear waste, mill tailings (with typical activity levels on the order of a thousand picocuries per gram) are still radioactive at levels a thousand times above natural background. In addition, mill tailings are produced in huge quantities and, as indicated by the 80,000- year half-life of thorium-230, their radioactivity is extremely long-lived. As of the beginning of 1990, a total of over 230 million metric tons of uranium mill tailings had accumulated in the U.S. from uranium production for nuclear weapons production and nuclear power generation. [25] This amount represents over 95 percent of the volume of all nuclear waste generated in the U.S. to date, [26] and presents a potential health hazard to communities nearby that will last for hundreds of thousands of years. Thus, if not managed properly, mill tailings, though low in level of radioactivity, have the potential to be the dominant contributor to radiation exposure from the nuclear power fuel cycle because of their extreme longevity. [27]

Mill tailings pose a number of health hazards—both because of their radioactivity and because they contain a variety of other non-radio-active but chemically toxic constituents (including remnants of the chemical solvents and leaching agents used to extract the uranium, as well as concentrations of heavy metals like arsenic, selenium, and molybdenum which are also contained in the ore along with uranium). [28]

One of the principal problems of concern is ill health effects due to the inhalation or ingestion of radionuclides from the tailings or of radon gas from the decay of these nuclides. Radon is continuously emitted from exposed mill tailings piles as a consequence of the radioactive decay of the radium and thorium in the tailings. Because of the long half-life of thorium-230 (80,000 years), the hazard from it and from the radon it produces will decline at a rate of only about one percent per thousand years. [29]

---

25. DOE 1990d, pp. 131, 148. This includes 188 million metric tons at licensed sites plus 44.5 million metric tons of mill tailings at inactive and DOE sites.
26. Based on the total volume of all U.S. radioactive waste (including spent fuel, high-level waste, transuranic waste, low-level waste, and uranium mill tailings) from all sources (both commercial and military) produced in the U.S. since the 1940s. Based on U.S. Department of Energy records as compiled in Saleska 1989, Appendix C.
27. As noted by Victor Gilinsky, Commissioner, U.S. Nuclear Regulatory Commission, in "Remarks Presented at the Pacific Southwest Minerals and Energy Conference," Anaheim, CA, May 2, 1978.
28. NAS 1986, p. 1.
29. *Ibid.*, p. 12.

Low-Level Radioactive Wastes

Commercial low-level radioactive waste (LLW) in the U.S. is defined by what it is not. According to Nuclear Regulatory Commission (NRC) regulations, low-level waste is "radioactive waste not classified as high-level radioactive waste, transuranic waste, spent nuclear fuel, or by-product material [i.e., uranium or thorium mill tailings]..."[30]

The "low-level" radioactive waste category thus includes everything from slightly radioactive trash (such as mops, gloves, and booties) to highly radioactive activated metals from inside nuclear reactors. It includes both short-lived and long-lived radionuclides.

More than 20,000 licenses have been issued by the NRC and NRC agreement states for the commercial handling and use of radioactive materials,[31] and most of these users will generate some sort of waste that will be disposed of as low-level waste. These 20,000-plus users include a diversity of sources, such as radiochemical manufacturers, research laboratories, hospitals, medical schools, universities, and non-DOE government agencies. The single largest source of commercial low-level waste, however, is the nuclear power industry, whose hundred or so licensees accounted for over 70 percent of the volume and almost 95 percent of the radioactivity of all low-level waste shipped for disposal in 1989.[32]

---

30. NRC 1988b (10 CFR Part 61.2).
31. The figure is from DOE 1990d, p. 95. The NRC has a program to share nuclear regulatory enforcement authority with states, as long as state regulations are not inconsistent with NRC regulations. States which participate in this program are referred to as NRC Agreement states.
32. DOE 1990d. Calculated from Table 4.2 (p. 106), and Tables 4.14 through 4.19 (pp. 118-123). Nuclear power industry low-level waste includes low-level waste disposed of from operating nuclear reactors as well as from nuclear fuel preparation activities such as uranium conversion and fuel fabrication. It does not include waste from uranium enrichment, which is conducted by the DOE and is disposed of as DOE low-level waste. However, since uranium enrichment wastes are estimated to represent only 3 to 4 percent of the total volume of low-level waste generated by nuclear fuel preparation activities, this exclusion does not alter the nuclear power industry contribution significantly.

Commercial Low-Level Waste from the Nuclear Fuel Cycle

All wastes generated by the commercial nuclear fuel cycle, aside from the uranium mill tailings and the spent fuel, are classified as "low-level" wastes. This includes wastes from the fuel preparation activities —conversion, enrichment, and fuel fabrication—as well as wastes (other than the spent fuel) from the operation of nuclear reactors.

The bulk of low-level waste from the nuclear fuel cycle (about 80 percent in 1989) is generated by the reactors themselves. There are two primary processes which underlie the generation of low-level waste at commercial reactors: microscopic leakage of fission products from fuel rods, and the creation of activation products as a result of neutron bombardment of non-radioactive elements inside the reactor.[33]

A small fraction—on the order of 0.1 percent—of the fuel rods inside a reactor suffer from microscopic pinhole leaks.[34] This allows some of the fission products which build up inside the fuel rods to escape into the cooling water surrounding them, contributing to its radioactivity. This cooling water is continuously being cleaned of radioactivity by ion exchange resins and demineralizer filtration units, which themselves become contaminated and are eventually disposed of as low-level waste. These spent resins and filters can be highly radioactive, and are sometimes capable of delivering a radiation dose exceeding 1,000 rems per hour, essentially a lethal dose.[35] The inevitable occasional leaks and spills of the radioactive cooling water give rise to maintenance and clean-up activities which result in further waste generation. For example, the mops, protective clothing worn by workers, detergents, and any other materials used to clean up a spill of radioactive water (such as evaporator solids used to process it), become contaminated and must be disposed of as low-level waste.

---

33. "Activation products" are created when a neutron from the nuclear fission reaction is absorbed by a non-radioactive atom. That atom can then become "activated" or radioactive. For example, when a common nitrogen-14 atom absorbs a neutron, it emits a proton and becomes radioactive carbon-14; stable cobalt-59, the most common form of cobalt found in nature and a constituent of stainless steel, can absorb a neutron and become radioactive cobalt-60.
34. As discussed in Carter 1987, p. 19.
35. Carter 1987, p. 19.

The other principal process contributing to low-level waste generation is the build-up of activation products as a result of the high
neutron flux in a reactor core. Corrosion products in the cooling water
(from the slight but incessant corrosion of the exposed metal surfaces
inside the cooling system) can become activated, mixing with the
fission products and contributing to the radioactivity of the cooling
water. In addition, the continuous neutron bombardment of the reactor
components makes these radioactive. Items such as fuel channels,
control-rod channels, or instrumentation placed inside the reactor core
are removed from the core from time to time and must be disposed of
as waste.

Table 2 shows the average amounts of low-level waste generated
at each stage of the nuclear fuel cycle which result from operating a
typical 1000 MW(e) commercial reactor for one year. This includes
waste generated at conversion, enrichment, and fabrication facilities, as
well as at typical pressurized water and boiling water reactors.

Table 2
LLW GENERATED TO SUPPORT TYPICAL
1000 MW(e) REACTOR* FOR ONE YEAR

	Volume (cubic meters)	Radioactivity (curies)
Fuel Preparation		
Uranium Conversion	6.7	75
Uranium Enrichment	2.3	85
Fuel Fabrication	58.3	0.14
Pressurized Water Reactor (PWR)		
Routine Waste	127.5	703
Irradiated Components	2.9	1,366
Boiling Water Reactor (BWR)		
Routine Waste	355.8	1,848
Irradiated Components	12.9	9,041

Source: DOE 1990d.
NOTE: *Assumes a reactor capacity factor of 65 percent.

There are six commercial disposal sites for "low-level" wastes in the U.S. where large amounts of radioactive waste have been disposed of by a method referred to as "shallow-land burial," in which wastes are stored in drums (and in the past, cardboard boxes) and buried in trenches. This primarily includes waste from nuclear power plants, but also from other industrial, medical, and research sources as well.

So far, three of the six disposal sites have been closed, and have experienced environmental problems. Table 3, below, shows the amounts of waste at each of these sites, according to DOE records.

Table 3 COMMERCIAL LOW-LEVEL WASTE DISPOSAL THROUGH 1989, BY SITE		
Disposal Site	**Accumulated Volume** (cubic meters)	**Radioactivity** (1000s of decayed curies)
Active		
Beatty, Nevada	116,657	518
Richland, Washington	318,258	1,995
Barnwell, South Carolina	616,022	4,646
Shut Down		
West Valley, New York	77,074	1,267
Maxey Flats, Kentucky	135,280	2,401
Sheffield, Illinois	88,334	60
Rounded Totals	1,352,000	10,885
Source: DOE 1990d, pp. 114-115		

Low-Level Waste Categories

NRC regulations sub-divide commercial low-level waste into four classes which are determined by the types of radionuclides and their concentrations which make up the waste. These classes are labeled Class A, Class B, Class C, and Greater-than-class-C. Class A waste is the least radioactive on average, and is contaminated primarily by what the NRC terms "short-lived" radionuclides. Classes B and C are more radioactive: Class B may be contaminated with greater amounts of "short lived" radionuclides than Class A, and Class C with greater amounts of long-lived and short-lived radionuclides than Class A or B. Greater-than-class-C waste is much more radioactive than the other classes, and generally is considered unacceptable for near-surface disposal, which is how Classes A, B, and C are generally disposed of in the U.S. Shallow-land disposal used to be simple dumps mainly, but the concept now also includes more elaborate structures.

Table 4 shows the average radioactivity characteristics of the various classes of waste.

Table 4
LOW-LEVEL WASTE CHARACTERISTICS

	Average Concentration (Curies/ft^3)	Selected Samples (Curies/ft^3)	
Low-Level Waste:			
Class A	0.1		
Class B	2	4.4	(NY Cintichem)
Class C	7	160	(NY reactor avg.)
Greater-than-C	300 to 2,500*		
Transuranic Waste:			
Contact-handled	0.57		
Remote-handled	47		
Military High-Level Waste:	100	920	(Savannah River sludge)
		3.7	(Hanford salt cake)
		7,110	(SRP Glass, proj.)
Commercial Spent Fuel	73,650**		

Sources: OTA 1988; DOE 1990d; and NYSERDA 1990.
* The 300 figure is based on the 1985 inventory. The higher figure represents anticipated inventory in 2020, including some decommissioning wastes.
** Based on average activity in all spent fuel at the end of 1989 and on overall fuel assembly dimensions.

As can be seen, some samples of Class C and even Class B wastes (e.g. 4.4 curies/ft^3 from New York's Cintichem facility) are more radioactive than some portions of high-level waste (e.g., salt cake in the Hanford waste tanks). The average Class C wastes from New York State's nuclear reactors are considerably more radioactive than military high-level wastes. The average radioactivity concentration in current Greater-than-class-C low-level wastes is about three times that of military high-level wastes.

Military Low-Level Waste

Low-level radioactive wastes are also produced as the result of activities associated with nuclear weapons research, design, materials fabrication, and assembly. Most of these wastes are currently buried in shallow trenches at DOE sites.

Prior to 1970, some low-level military wastes (including those contaminated with transuranics) were dumped at sea. In total, according to DOE records, nearly 90,000 containers, originally containing some 94,000 curies of radioactivity were dumped at a number of sites off both the east and west coasts.[36] In 1977, the U.S. Environmental Protection Agency estimated that as many as 25 percent of the containers at one ocean dump were leaking.[37] Eventually, of course, as a result of the inevitable corrosion by seawater, they will all leak.

Table 5
MILITARY LOW-LEVEL WASTE DISPOSAL, BY SITE

Disposal Site	Annual Disposal Volume (1987-1989 average, m^3)	Total Accumulation (through 1989, in m^3)
Principal Sites		
Los Alamos National Lab	4,800	205,400
Idaho National Engineering Lab	2,100	143,100
Nevada Test Site	19,500	279,900
Oak Ridge National Lab	800	206,900
Hanford	16,400	563,800
Savannah River	32,500	598,800
Other Sites		
Femald Plant	0	298,500
Oak Ridge Y-12 Plant	10,800	146,500
All Others	1,400	114,100
TOTAL VOLUME	88,300	2,557,000
TOTAL RADIOACTIVITY (decayed curies)	1,371,000	13,771,000

Source: DOE 1990d, pp. 105, 107, 108.

---

36. DOE 1990d, p. 109.
37. Nucleonics 1977, p. 12.

During the 1980s, the DOE generated between 60,000 and 120,000 cubic meters containing one to two million curies of low-level waste per year. Most of this has been disposed of by burial at six principal sites. Table 5 lists the rates and total accumulations of military low-level waste at major DOE sites through the beginning of 1990. Figure 5 depicts the amount on a map of the U.S. which shows roughly where they are located. Also on this map are indicated the sites where ocean-dumping of military low-level waste occurred.

Figure 5
LOCATIONS* AND VOLUMES OF MILITARY LOW-LEVEL WASTE

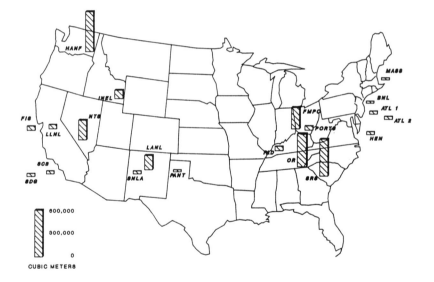

Source: DOE 1990d, p. 98.
* Land-based Sites: HANF=Hanford Nuclear Reservation (Richland, Washington); LLNL=Lawrence Livermore National Laboratory (Livermore, California); NTS=Nevada Test Site (Mercury, Nevada); INEL=Idaho National Engineering Laboratory (Idaho Falls, Idaho); SNLA=Sandia National Laboratory (Albuquerque, New Mexico); LANL= Los Alamos National Laboratory (Los Alamos, New Mexico); PANT=Pantex Plant (Amarillo, Texas); PAD=Paducah Gaseous Diffusion Plant (Paducah, Kentucky); FMPC=Feed Materials Production Center (Fernald, Ohio); PORTS=Portsmouth Gaseous Diffusion Plant (Portsmouth, Ohio); OR=Oak Ridge Complex (Oak Ridge, Tennessee);SRS=Savannah River Site (Aiken, South Carolina); BNL=Brookhaven National Laboratory (Brookhaven, New York). West Coast Ocean-Dump Sites: FIS=Farallon Islands (Central California Coast); SDG=Off-Coast (San Diego, California); SCB=Santa Cruz Basin. East Coast Ocean-Dump Sites: Mass=Massachusetts Bay; ATL=Atlantic Site 1 (38'31'N, 72'06'W); ATL 2=Atlantic Site 2 (37'50'N, 70'35'W); HEN=Cape Henry (Virginia Coast).

Radioactive Wastes from Decommissioning Nuclear Reactors

At the end of its useful lifetime, a reactor or other nuclear facility must be shut down, cleansed of radioactivity to the extent possible, and dismantled or entombed. This process is known as decontamination and decommissioning. No large commercial reactor has yet undergone this shut-down process. [38] However, the process is expected to involve a number of steps, including flushing out the cooling system with chemical chelating agents to remove radioactive residues, washing equipment, and the scrubbing, chipping, and sandblasting of walls and floors. This is intended to remove superficial radioactivity. Metal components within the reactor will have to be cut up with a cutting torch operated remotely to protect workers from the intense radiation. [39]

Three main decommissioning scenarios are usually considered:

- Entombment. Under this scenario, a reactor would be shut down and decontaminated to a superficial degree, then sealed off with concrete and fences. Numerous huge reactor hulks, contaminated with long-lived radionuclides, would be left where they stand for many generations to monitor and maintain.

- Mothballing. Referred to as "SAFSTOR", this is deferred decommissioning. The reactor would be sealed for 30 to 100 years to permit some of the shorter-lived radionuclides to decay, after which the dismantling would be completed.

- Immediate Decommissioning. (Called "DECON") This is the mode currently assumed by DOE waste projections, and what most utilities say they are planning to use. Under this scenario, decommissioning would take place in the space of about six years, beginning two years after shut down.

38. A number of small reactors have been decommissioned. The largest of these, however, is the 72 megawatt (electric) Shippingport reactor in Pennsylvania. This is less than one-tenth the size of a typical large modern commercial reactor, which can be 1,000 megawatts (electric) or more. No reactor greater than 500 MW(e) is scheduled to be decommissioned before 2004 (DOE 1990d).
39. Carter 1987, p. 22.

Most utilities currently appear to be planning immediate decommissioning. According to a survey of utility plans, 105 of 124 reactors considered were slated for immediate decommissioning. [40]

Huge amounts of waste will result from decommissioning of these nuclear power reactors, which will not occur until some time in the next century.

According to DOE's official projections (which are based on the immediate decommissioning scenario), nearly 40 percent of the total volume of low-level waste produced by a boiling water reactor over its anticipated 40-year lifetime is from decommissioning. For PWRs, which produce less operating low-level waste than do BWRs, but about the same amount of decommissioning waste, almost two-thirds of lifetime waste comes from decommissioning. See Figure 6.

However, as Table 6 shows, if decommissioning reactors is delayed for 30 to 100 years, radioactive decay will result in a decrease in the amount of radioactivity in the waste by about 20 to 50 times for a PWR, and 35 to 70 times for a BWR. As the table also shows, delay is expected to result in a reduction of decommissioning waste volume for delays of 50 years or more, mostly due to a reduction in the amount of Class A waste. A delay of only 30 years will not result in significant waste volume reduction, but at 50 years, the decommissioning waste volume is reduced by a factor of 10. Waiting 100 years continues to reduce activity, also resulting in slight additional volume reductions. [41]

---

40. Borson 1990, pp. 87-88.
41. DOE 1990d, p. 174. Note that reductions in radioactivity are not always accompanied by equivalent reductions in waste volume. This could be due to the presence of different radionuclides with varying half-lives in a given quantity of waste. A given period of time could result in significant reductions in activity due to the decay of short-lived radionuclides, while the presence of the long-lived radionuclides still requires nearly the entire volume to be considered radioactive waste.

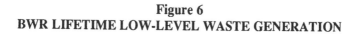

Figure 6
BWR LIFETIME LOW-LEVEL WASTE GENERATION

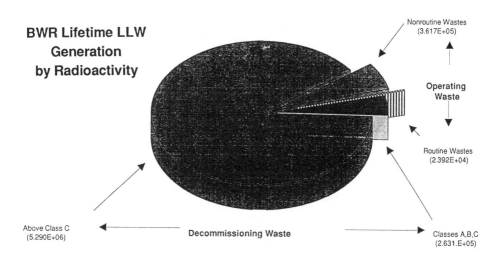

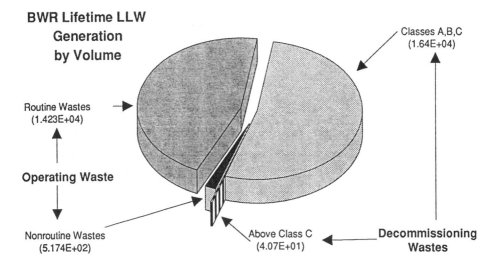

Source: Adapted from data in DOE 1990d, p. 223.

Table 6
ESTIMATED VOLUMES AND ACTIVITIES OF WASTES FROM DECOMMISSIONING ALTERNATIVES CONSIDERED FOR REFERENCE LIGHT WATER REACTORS[a,b]

Decommissioning Alternative	Totals Volume (m³)	Totals Activity (10³ Ci)	Class - A LLW Volume (m³)	Class - A LLW Activity (10³ Ci)	Class - B LLW Volume (m³)	Class - B LLW Activity (10³ Ci)	Class - C LLW Volume (m³)	Class - C LLW Activity (10³ Ci)	Exceeds Class - C LLW limits Volume (m³)	Exceeds Class - C LLW limits Activity (10³ Ci)
Reference boiling - water reactor [1,155 MW(e)]										
Immediate decontamination following shutdown	18,985	6,595.8	18,512	13.9	373	42.8	53	239.1	47[c]	6,300.0
Deferred decontamination after a safe storage period of:										
30 years[d]	18,985	180.4	18,652	1.4	233	1.1	53	6.5	47[c]	171.4
50 years[d]	1,783	141.4	1,450	0.2	247	1.0	39	4.7	47[c]	135.5
100 years[d]	1,673	97.2	1,340	0.1	27	0.6	39	3.3	47[c]	93.2
Entombment[e]	8,078	6,586.6	7,605	4.7	373	42.8	53	239.1	47[c]	6,300.0
Reference pressurized - water reactor [1,175 MW(e)]										
Immediate decontamination following shutdown	18,325	4,906.2	17,961	37.3	214	53.1	17	34.3	133[f]	4,781.5
Deferred decontamination after a safe storage period of:										
30 years[d]	18,328	209.1	18,055	1.5	123	0.6	17	1.5	133[f]	205.5
50 years[d]	1,833	159.4	1,568	0.3	115	0.2	17	1.1	133[f]	157.8
100 years[d]	1,783	106.2	1,533	0.2	100	<0.1	17	0.8	133[f]	105.2
Entombment[e]	3,500	4,908.0	3,136	39.1	214	53.1	17	34.3	133[f]	4,781.5

Source: DOE, 1990d, p. 174.

[a] Data for each reactor are based on 40 years of opeariion and a capacity factor of 0.75.
[b] Based on a limiting concentration of long- and short - lived radionuclides given in Tables 1 and 2 of 10 CFR 61.55.
[c] Contribution from the core shroud.
[d] Includes radioactive wastes from both preparations for safe storage and deferred decontamination.
[e] Involves the removal of reactor spent fuel (shipped to repository) followed by the encasement of the rest of the radioactive portion of the reactor facility.
[f] Contributions from the lower core barrel, thermal shields, lower grid plate, and core shroud.

Other Wastes

Mixed Wastes

In both the commercial and military sectors, some of the radioactive wastes generated are mixed with hazardous substances, such as organic solvents or other toxic chemicals. This is a problem for low-level waste in the commercial sector, as well as for low-level and transuranic wastes generated for military purposes by the DOE. Much of this waste (especially the transuranic waste) contains substantial quantities of long-lived radionuclides, such as plutonium-239 and technetium-99.

The radioactive components of mixed wastes are regulated under the Atomic Energy Act by the NRC for commercial sources, and by the DOE for military sources. The hazardous components, however, are subject to regulation by the EPA according to an environmental law known as the Resource Conservation and Recovery Act (RCRA). Complying with RCRA is proving difficult at nuclear facilities, especially those run by the DOE, in part because until 1987, the DOE claimed it was exempt from the law, and its facilities and procedures were not run to take the law's provisions into account.[42] It was only recently that the DOE even began to compile a comprehensive complex-wide database of mixed low-level waste of the sort it has maintained for the various categories of radioactive waste.

Mixed waste at DOE sites includes a broad range of materials, such as cleaning solutions and cleanup materials, engine oils and grease, contaminated soils, building materials, water treatment materials, and decommissioned weapons manufacturing equipment.[43] The DOE has so far identified a total inventory of about 56,000 cubic meters of mixed low-level waste, and is generating new mixed low-level waste at a rate of about 2,500 to 8,000 cubic meters a year.[44]

Mixed waste at commercial power plants includes radioactively contaminated organic chemicals, waste oil, CFC solvents and concentrates, chromium wastes from resin changeouts at reactors, cadmium wastes from spent reactor equipment and cleanup activities, lead solu-

---

42. On May 1, 1987, in part as a result of legal pressure brought on by environmental groups, the DOE published an interpretive ruling (DOE 1987, p. 15937), which clarified that DOE hazardous wastes were subject to RCRA.
43. DOE 1990d, p. 187.
44. DOE 1990d, pp. 194, 196, and 201.

tions from lead shielding decontamination, aqueous corrosive liquids from cleaning spent fuel casks and resin filters, etc. [45] An estimated 3 to 10 percent of the total volume of commercial low-level waste from all sources is mixed waste. According to one source, a typical U.S. commercial reactor generates on the order of 100 cubic meters of mixed low-level wastes each year.[46]

NARM Wastes

NARM wastes (Naturally-occurring and Accelerator-produced Radioactive Materials) are orphan wastes not consistently regulated under any current federal standard. NARM includes such materials as radium-226 and thorium-230 produced outside the nuclear fuel-cycle, and radionuclides produced by particle accelerators. NARM wastes are generated by both federal and non-federal facilities. [47]

The largest volume of NARM wastes consists of naturally occurring radioactive material such as radium, thorium, and uranium where they occur outside of explicitly nuclear fuel cycle activities such as uranium milling (where they would be regulated as uranium mill tailings). One example of this sort of waste are those left behind from the processing of uranium ores for the extraction of radium. Such activities amounted to a sizable industry in the early years of this century, when radium was for a time a valuable commodity.

---

45. DOE 1990d, pp. 188, 205.
46. Nuclear Management and Resources Council, *The Management of Mixed Low-Level Radioactive Wastes in the Nuclear Power Industry,* NUMARC/NESP-006, January 1990, as cited in DOE 1990d, p. 205.
47. EPA 1990b.

Chapter 3

OVERVIEW AND CRITIQUE OF THE CURRENT APPROACH TO RADIOACTIVE WASTE MANAGEMENT

Generally, the long-term management of mill tailings, low-level waste, transuranic waste, high-level waste, and spent fuel have been considered as separate issues, with separate solutions.

Current plans call for the disposal of low-level waste at various sites around the country; some transuranic wastes in a geologic repository near Carlsbad, New Mexico (The Waste Isolation Pilot Plant known by its acronym WIPP); and spent fuel and high-level radioactive waste in a separate geologic repository which has yet to be built. Uranium mill tailings are generally being dealt with on-site or near the site, according to Environmental Protection Agency (EPA) standards.[1] Considerable problems remain to be addressed in each of these areas. There are no firm plans—in some cases not even clear, viable proposals—for large quantities of these wastes, most notably in the weapons complex, which is just starting to deal with the vast environmental contamination problems it has created over the past 45 years. Cost estimates for the clean-up run upwards of $150 billion, notwithstanding the fact that some sites may be irreversibly contaminated.

The main focus of our analysis here is on the long-lived and more dangerous components of the conventional waste categories. This

---

1. EPA 1983c, a (40 CFR 192).

includes:

- Commercial spent fuel, high-level reprocessing waste, and transuranic waste, all of which (except for part of the transuranic waste) according to current U.S. law and policy are slated for permanent disposal in a deep geologic repository.

- The long-lived or especially dangerous components of low-level waste from both commercial and military sources, most of which are currently disposed of through shallow burial in radioactive waste landfills.

- Uranium mill tailings, especially concerning the long-lived components radium-226 and thorium-230.

In the sections below, we consider U.S. plans for dealing with each of these issues.

A. Spent Fuel, High-Level, and Transuranic Wastes

Background on Nuclear Waste Disposal Issues

Historical Overview

As discussed in Chapter 1, one of the major problems associated with radioactive waste is the fact that much of it will be radioactive—and thus will require isolation from the human environment—for hundreds of thousands, if not millions, of years. Since this is a time period far longer than all of recorded history, the problem of waste disposal presents an enormous challenge.

Yet when the first high-level nuclear wastes were produced in the 1940s as a result of the Manhattan Project to construct the first atomic bomb, they were stored in what were at the time considered to be "temporary" storage tanks. There was no plan for permanent disposal, and as far as the U.S. government was concerned, the exceptional and pressing circumstances of World War II relegated such long-term issues to a low priority.

Over ten years later, when the exigencies and shortages of that war were long past, the U.S. government made a definitive commitment to commercial nuclear energy by licensing the first commercial reactor in 1957—even though there was still no waste disposal solution in sight. In remarks before the National Academy of Sciences conference in 1955, for example, A.E. Gorman of the Atomic Energy Commission's (AEC) reactor development division acknowledged that the attitude of the Commission had been to "sweep the problem under the rug."[2] And as Carroll Wilson, the first general manager of the AEC, acknowledged much later,

> Chemists and chemical engineers were not interested in dealing with waste. It was not glamorous; there were no careers; it was messy; nobody got brownie points for caring about nuclear waste. The Atomic Energy Commission neglected the problem... The central point is that there was no real interest or profit in dealing with the back end of the fuel cycle.[3]

Today, 50 years after the first sustained fission reaction, and over 30 years after the first commercial reactor began operating, a great many studies have been done at much expense, but the subject is still controversial and there is still no demonstrated long-term solution to the million-year disposal problem presented by nuclear waste.

The disposal of highly radioactive waste deep below the earth's land surface in mined geological repositories was the first form of disposal seriously proposed (in 1957),[4] and is today the legally designated (though not yet technically demonstrated) form of disposal. However, there have been many other alternatives proposed and researched over the years.

Among the options that have been considered are disposal by shooting into space, or emplacement under the Arctic ice cap. Although the land-based deep geologic repository approach is the legally designated solution, continuing research on sub-seabed disposal is also authorized. This is essentially disposal in geologic formations under the ocean floor with an emphasis on areas which appear to have long-term stability.[5]

---

2. Carter 1987, p. 54.
3. Wilson 1979, p. 15.
4. NAS 1957.
5. A number of these alternatives are discussed in DOE 1979. For an overview of

Current Law

In 1982, Congress passed the Nuclear Waste Policy Act, a law which designated deep geologic disposal as the preferred technical solution, essentially curtailing or terminating serious research and development on other disposal methods. Revisions to the law in 1987 established an Office of Sub-seabed Disposal to explore the potential of the geologic layers under the ocean floor for containing wastes, although this office has had little support from the DOE.

The law provided states selected as potential hosts for the repositories with a limited veto and recognized the rights of Native American tribes as equal to the states. It finally—25 years after the advent of commercial nuclear power—provided for industry financing of waste disposal through the establishment of a Nuclear Waste Fund (funded so far by a 0.1 cent per kilowatt-hour fee imposed on nuclear electricity). The law also obligates the DOE to contribute to this fund for DOE's repository-destined military wastes. The DOE has contributed only $5 million to this fund so far, an insignificant sum even when compared to the level of obligation of about $500 million which the DOE itself acknowledges.[6,7] It also imposed the requirement that explicit site selection and environmental criteria be adopted, and that a final site be selected from among numerous sites examined on the basis of detailed characterization studies.

However, DOE's problem-ridden site selection process, flaws in the 1982 law and in EPA and NRC regulations (see below), and vigorous citizen opposition led the DOE and others once more to a more politically convenient solution.[8] Congress amended the Nuclear Waste Policy Act in 1987,[9] overriding many of the site selection and characterization provisions of the 1982 act and designating Yucca Mountain, near the U.S. Nuclear Test Site in Nevada, as the sole site to be examined as a candidate for the first high-level waste repository. Thus, the final selection of Nevada came about as a result of a process that, in part,

---

sub-seabed disposal see OTA 1986, or Hollister 1981.

6. Personal communication from Ron Callen, to Arjun Makhijani (11 June 1991).
7. DOE 1989a.
8. The DOE's problems in implementing the 1982 act are discussed in more detail in Makhijani 1989, pp. 37-43.
9. The 1987 Nuclear Waste Policy Amendments Act, Public Law 100-203.

started with the fact that the government already controlled the land, and ended in a decision in which politics overwhelmed science.

One misguided element that was a prominent part of both the 1982 and 1987 legislation was a sense of urgency about the need to develop a permanent solution for nuclear waste as soon as possible. The lack of a disposal solution had long been a political albatross around the neck of the nuclear industry, and so industry lobbyists exerted strong pressure to get the job done quickly. As one observer of the 1982 legislation noted, "the industry feeling was that the sooner this solution could be effectively demonstrated, the better for the industry politically."[10]

Some anti-nuclear, environmental, and arms control groups, for their part, wanted to head off the reprocessing option for commercial spent fuel by disposing of it as soon as possible. The 1982 law thus established target dates for repository siting and construction that were far too ambitious. These target dates came to be seen as deadlines which only increased the sense of urgency about the waste disposal problem and, when these deadlines were not met, DOE's already low credibility was even further eroded.

Current Standards and Regulations for Long-Term Management

The extreme longevity (on the order of millions of years) of some of the radionuclides in high-level waste means that it is impossible to guarantee that this waste will remain completely isolated from the environment. "Isolation," then, becomes a relative term in which it is assumed that some radioactivity will be released into the environment over time. In fact, according to the National Academy of Sciences, "[e]ssentially all of the iodine-129 [half-life: 15.7 million years] in the unreprocessed spent fuel in wet-rock repositories will eventually reach the biosphere."[11] As the EPA has remarked, any environmental standards regulating "acceptable" releases of radioactivity from nuclear waste repositories must therefore "address a time frame without precedent in environmental regulations."[12]

Proposed standards are thus all based on the assumption that some

10. Carter 1987, p. 196.
11. NAS 1983, p. 11.
12. EPA 1985, p. 38066.

radioactivity from a repository will reach the environment. Given this assumption, there are several possible approaches to developing standards.

One approach would be to establish upper limits on the "acceptable" health risks for both individuals and the population as a whole.[13] An acceptable health risk can be translated into a radiation dose limit, which can be put together with assumptions about environmental transport of radioactivity in order to establish repository performance standards. The performance standards can then be used to establish selection guidelines to evaluate the suitability of potential repository sites. This results in what is sometimes called "health-based" (or "risk-based") standards.

Another approach results in "technology-based" standards. This approach starts by considering what is possible, based on the current state of scientific knowledge, the best currently available technology, and a reasonably good geological site. Technology- based performance standards can then be developed which are intended to ensure that a good site is chosen and that the technology used to construct a repository is appropriate. They are oriented towards assuring that compliance with the standards can be measured and achieved. This can be done *independent* of the health risks that would arise from the standards that

---

13. It is important here to understand the concepts of individual and population risks. An individual risk is often expressed in terms of the chance that an individual might get a fatal cancer (for example, a typical risk limit for individuals in environmental regulation is one in one million). A population risk is often expressed in number of expected deaths. Although the risk to individuals is important to consider, the risk to the population as a whole is just as meaningful from a public policy perspective. This is because the total population risk is what indicates how many people will die as a result of a given risk. Thus, population risk provides a basis for arguing that a "small" individual risk of death imposed on a *large* number of people is morally worse than the same-sized individual risk imposed on a *small* number of people. The risk to each exposed individual is the same in both cases, but in the first case, a larger number of people will die. Conversely, even a small overall population risk, if borne primarily by a small number of people, can result in huge risks to each of the few individuals exposed. For example, if an environmental risk sufficient to kill one person is spread evenly among 10 people, each will have a one in 10 risk of dying. Spread among one million people, each individual will have a one in a million risk. In each case, the population risk is the same (one death), but we would generally consider the risks to the arbitrarily exposed individuals to be unacceptably high in the former case. A humane public health and environmental policy will minimize both types of risk, individual and population.

result, which might be either large or small.

We believe that the ideal route to environmental protection would rely on the first approach, which takes the health of human beings as its primary standard. The second approach essentially relies on circular reasoning: standards are set so as to ensure that the standards can be met. When taken to its extreme, this represents an attitude that environmental contamination is an inevitable result of technological progress, that human beings have no choice but to live with the consequences, and that environmental regulation is limited simply to a kind of "sweeping up" after the mess is made.

The first approach, however, as an across-the-board model, provides a much more complete conception of the possibilities for environmental preservation; comprehensively applied, it would extend health- and environmentally-based criteria to decision-making about which technologies are selected in the first place to fill a given societal need, and how manufacturing and production processes are designed.

Unfortunately, as we have discussed elsewhere, little thought went into the consequences during the years when nuclear technology was under development, and we are now confronted with a situation where the waste has been produced. Under such constrained circumstances, it is arguable that the second technology-based approach is as good an approach as can be taken. This is, in fact, the approach which the EPA took as its primary one in developing its standards.

Unfortunately, both of these approaches were essentially abandoned by federal law. The 1982 Nuclear Waste Policy Act asked the DOE to submit a list of potentially acceptable repository sites within six months of promulgation of the law. It did not require the EPA, however, to come up with final standards for acceptable dose limits—logically the first step—until six months after the DOE's initial repository list was submitted. Further, the NRC was to establish technical performance standards (theoretically consistent with the EPA standards) for the repository about the same time that the EPA standards were supposed to be issued.

The way the process has worked in practice is even worse than provided for in the provisions of the 1982 waste policy law. The EPA's final standards were delayed, and therefore were not issued until 1985 over two years after the final NRC standards.[14] Further, the EPA

---

14. EPA standards were published on 19 September 1985. The NRC standards were

standards were challenged in court by several states and environmental groups on the grounds that, among other things, they were too lax, and violated provisions of the Safe Drinking Water Act. The court agreed that the standards violated the Safe Drinking Water Act, and vacated them, requiring the EPA either to bring the standards into compliance with existing law, or explain why the inconsistency exists.

The EPA is now in the process of considering the issues raised by the court which vacated the standards, and plans to issue new standards in 1992. Because the new standards are likely to be similar in many respects to the 1985 standards, we will review the vacated EPA standards below along with the NRC regulations.

1985 EPA Standards

The EPA environmental standards for waste repositories[15] set out long-term containment requirements that limit releases of radioactivity to the environment for 10,000 years. These are expressed in terms of radionuclide release limits for the rate of release of individual nuclides in the waste to the environment.

Corresponding to the radionuclide release limits is an overall population dose limit extending to 10,000 years; individual doses are limited for only 1,000 years. The overall population dose from all radioactivity escaping from the repository may not cause more than 1,000 premature cancer deaths over the 10,000-year period (or, one extra fatal cancer, on average, every 10 years).[16] The individual dose to any member of the public from radioactivity escaping from the repository is limited to no more than 25 millirems per year.[17] After the first 1,000 years, only the overall population dose limit remains, and there is no cap on individual doses.

There are also groundwater protection requirements which limit the concentrations of radioactivity from the repository in groundwater to such that persons who draw all of their drinking water from that

---

published on 21 June 1983.
15. The EPA standards were incorporated into federal regulations at 40 CFR 191 (EPA 1985).
16. EPA 1985, p. 38069.
17. EPA 1985, p. 38068. This is the same as the dose limit allowed to individual members of the public from the nuclear power fuel cycle by EPA standards at 40 CFR Part 190.

source of groundwater would receive not more than four millirems per year. Like the individual dose limits, these standards apply only for a period of 1,000 years after disposal.

As pointed out in a 1983 report by a panel of the National Academy of Sciences (NAS) (which criticized the draft EPA standards two years before the final ones were issued),[18] as well as by the EPA itself,[19] the 1,000- and 10,000-year time frames are short, given the long-term risk posed by some radionuclides. The NAS panel said it was possible for groundwater to become contaminated long after 10,000 years, and that the population dose received after this time could actually increase beyond the EPA limits, even if these limits were not exceeded within the first 10,000 years. The panel pointed out that some long-lived radionuclides would be capable of delivering very high doses from groundwater for many tens of thousands of years.[20] For example, according to one set of assumptions about radionuclide release from reprocessing waste made by the National Academy of Sciences, significant doses from the radionuclide lead-210 could peak shortly before 100,000 years, and continue for long afterwards. The possible doses from cesium-135 peak at almost one million years.[21]

Another problem with the 1985 EPA standards is that after 1,000 years, only overall population dose is limited; there is no limit after this time on the maximum dose an individual might receive.[22] This should be of particular concern given that the NAS panel found that under certain conditions radionuclides from a repository could deliver annual individual doses from neptunium-237 of up to 10,000 rems—a lethal dose.[23] According to the1985 EPA standards, this could happen without any violation of the standards, as long as the total number of people exposed was within the population limit.

In its study, the NAS panel proposed a slightly different basis for standards. Rather than limiting overall population exposure and individual exposures for only 1,000 years, the Panel advocated an individual dose-based standard for all future times:

---

18. NAS 1983.
19. EPA 1985, p. 38076.
20. NAS 1983, pp. 227-228.
21. NAS 1983, Figure 9-1, p. 256.
22. Although, as we note below, the new EPA standards may extend the dose limit to 10,000 years.
23. NAS 1983, p. 225.

> [T]he most meaningful and useful form of the criterion is the annual or lifetime radiation dose to an individual exposed at some future time to radionuclides released to the environment from a geologic repository. We have adopted as our criterion an annual dose of 10^{-4} Sv [10 millirem] to an individual, averaged over his lifetime, calculated at all future times.[24]

Missing from this recommendation, however, is the concept of a population dose limit. With the addition of a population dose limit, such an approach is a sound one, because it would place an upper limit to the amount of health risk to any individual from radioactive waste for all time, and it would also limit the total number of deaths as well.

The EPA is now moving further in this direction with its new draft standards, which retain the population dose limit, extend the individual dose limit to the entire 10,000-year time frame, and may also lower the individual dose limit to 10 millirems.[25] Whether these changes make it into the actual final rule remains to be seen.

However, it appears highly unlikely that the EPA will promulgate standards which address the criticisms pertaining to the shortness of the overall 10,000-year time frame.[26] The basic reason for this is that the current state of knowledge about geological processes and risk assessment simply does not allow confidence in predictions or control of the situation beyond this time frame. As it is, 10,000 years is probably pushing the limits of what is possible, given the current knowledge base.

This is an illustration of the fundamental problems posed by the whole nuclear waste dilemma. Nuclear waste will present a significant hazard for far longer than the 10,000-year regulatory time-horizon currently envisioned. Rather than address this defect, let alone that it was not even considered before proceeding with nuclear technology in the first place, current thinking simply accepts such shortcomings—indeed, even accepts dose-levels which may actually rise significantly after the 10,000 years, according to the NAS and the EPA—because we cannot currently do better.

---

24. NAS 1983, p. 212.
25. EPA 1991, pp. 18, 48.
26. As discussed in EPA 1989, and as is clear from the proposed new rule, EPA 1991.

NRC Standards

The Nuclear Regulatory Commission promulgated its high-level waste repository performance standards in 1983.[27] These standards require:

- complete containment of the waste within the waste packages for "not less than 300 years nor more than 1,000 years after permanent closure of the geologic repository."

- that after 1,000 years, the release rate of radioactivity "shall not exceed one part in 100,000 per year..." (based on the amount of radioactivity that is expected to be present at 1,000 years).[28]

- that the travel time of groundwater along the fastest likely path "shall be at least 1,000 years or such other travel time as may be approved or specified by the Commission."

- that the waste be retrievable from the repository during repository loading, and after repository closure until "significant uncertainties have been resolved, thereby providing greater assurance that the performance objectives [itemized above] will be met."

As mentioned above, these standards were supposed to be consistent with the EPA health standards—health standards which were not finalized until after these NRC standards were published. (Although it should be noted that the Nuclear Waste Policy Act requires the NRC standards to be amended, if necessary, to conform with the EPA standards.)

Thus, it can be seen that for a number of aspects of repository performance, federal standards as they currently exist are inadequate, and provide little assurance—especially over the long-term—that the

---

27. At 10 CFR Part 60.
28. This implies that the release of radioactivity from the repository is controlled by the NRC standards for about 100,000 years (after this time, 100,000 parts out of 100,000 — i.e. all the radionuclides — are allowed to have leaked).

environment or future public health will be protected for the length of time indicated by the characteristics of the radioactive materials.

Interim Management for Spent Fuel: MRS and Onsite Storage

Most of today's growing inventory (currently at approximately 24,000 metric tons) of spent fuel is stored onsite at the nuclear power reactor where it was produced in spent fuel storage pools; these pools were originally intended to be used for temporary cooling until the spent fuel was disposed of or reprocessed. Over the years, as plans for reprocessing were canceled, and long-term disposal options deferred, many of these at-reactor storage pools began to approach their capacity. As a result, many of these pools have been re-racked to hold more spent fuel than originally anticipated, and some of the later reactors have been constructed with pools designed to hold the reactor's anticipated lifetime inventory of spent fuel.

However, the spent fuel pools will not be capable of accommodating all spent fuel expected to be produced. The many delays (discussed further below) in the promised repository date have made it inevitable that additional onsite or other interim storage will be needed for some spent fuel from most nuclear reactors. The primary technology being considered today is the use of dry storage casks located either at a centralized Monitored Retrievable Storage (MRS) facility, or locally at each reactor site.

Centralized Away-From-Reactor Interim Storage at MRS

The 1982 Nuclear Waste Policy Act (NWPA) required the DOE to prepare a proposal for the construction of a Monitored Retrievable Storage (MRS) facility. The stated purpose of such an MRS facility would be to receive and prepare irradiated nuclear fuel from commercial reactors for temporary storage (in dry storage containers) before final disposal in a geologic repository.[29]

Part of the purpose of the MRS system was to allow the DOE to accept waste for storage up to eight years in advance of the opening of a permanent repository. It could therefore reduce the need for utilities

---

29. GAO 1986b, p. 4.

to expand temporary waste storage facilities at nuclear plant sites.[30]

This ability to reduce utility needs to build expanded storage—an entirely self-serving reason benefitting primarily nuclear utilities—is, however, as discussed below, the only clearly identifiable motivating reason to build an MRS in the system as currently envisioned.

Nonetheless, the 1987 amendments to the NWPA formally authorized the MRS, but also imposed a number of restrictions on its development which linked it to the opening of a permanent repository. These restrictions were the result of fears that an MRS could become a de facto permanent storage site. This concern has been one of the most consistently voiced criticisms of the MRS. The concern is that once it is constructed, an MRS is likely to become a substitute for a repository, thereby removing necessary governmental incentive for the development of long-term solutions to the nuclear waste problem.

In addition, there appears to be no other overall advantage to building an MRS. There is no cost advantage, and in fact there are recognized system cost *dis*advantages that are acknowledge by the DOE. The DOE's own cost estimates, over a range of scenarios and repository availability dates, indicate that an MRS will result in cost increases for all scenarios considered. These MRS-related cost increases range from $1.7 billion to $2.2 billion.[31] The MRS Review Commission's own study, mandated by the 1987 amendment to the 1982 Nuclear Waste Policy Act, which included consideration of increased costs for storage at shut-down reactors not considered by the DOE, found that in all scenarios considered but one, MRS would still increase nominal constant-dollar costs by up to $2.2 billion. In only one case, introducing an MRS in the overall waste management system was found to result in an $800 million savings, but even these savings were reversed in favor of a $500 million increase when the Review Commission discounted all costs to present value.[32]

Regarding risks due to transportation of spent fuel, the study by the MRS Review Commission found no significant difference between the MRS and No-MRS scenarios. However, this assessment is highly dependent on the location of the eventual repository with respect to the MRS because overall transportation requirements will depend on that.

30. GAO 1988a, p. 4.
31. DOE, cited in MRSRC 1989, p. 65. Costs are in 1989 dollars.
32. MRSRC 1989, p. 73. The MRS Review Commission used a discount rate of 4 percent. See footnote 19 in Chapter 4 for an explanation of cost discounting.

The Nuclear Waste Policy Act prohibits location of an MRS in the same state as a candidate repository site.

Extended onsite storage will actually considerably decrease transportation risk for two reasons. First, the quantities of cesium-137, strontium-90, and plutonium-241, among the most dangerous radionuclides, will greatly decrease due to radioactive decay, reducing accident consequences. Second, extended onsite storage will allow time for better casks to be developed, thereby further reducing risks of accidental radioactivity releases.

Regarding overall system risks, including occupational, public, and environmental effects, the Commission found that "the differences in risks among the alternatives considered are so small that they do not provide a basis for discriminating between MRS and No-MRS alternatives."[33]

Despite these issues, the DOE is still planning to build an MRS, and the administration is currently supporting energy policy legislation that includes an abolition of the schedule linkages between an MRS and a repository.

One of the driving incentives for the DOE appears to be a consequence of DOE's interpretation of the 1998 target date for repository opening contained in the 1982 law. The DOE used this date as a basis to enter into contracts with some utilities to take the spent fuel off their hands at this date. With the subsequent delay of the repository timetable by at least 12 years, this leaves the DOE with an obligation to take the spent fuel, but no place to put it, unless an MRS is built.

Demonstrating the frustration and exasperation of some utilities with the lack of progress and the delays in the waste program, a utility executive recently said at a DOE-sponsored meeting that he wanted the government to take charge of the spent fuel by 1998 and "I don't care where you put it."[34]

33. MRSRC 1989, pp. 52, 45.
34. Comments of a nuclear utility executive at the January 15-16, 1991 DOE-sponsored meeting on strategic principles for high-level waste management. The meeting was open to the public and the comments could be cited, but the ground rules were that the participants were not to be named.

Onsite Interim Storage in Dry Casks

Dry cask storage, which would be used at an MRS, can also be used as an alternative method (to the use of irradiated fuel pools) of storing irradiated fuel rods at reactor sites. It can be accomplished in various types of casks, modules, or vaults located outside the pools.[35]

Dry casks have several potential advantages over storage in water pools. First, since they do not contain water, which is necessary to enable a nuclear reaction in light water reactors, there is no chance of an accidental chain reaction, as there would be in water storage pools.[36] Second, since there is no water circulation and filtering, no "low-level" radioactive waste is produced by fuel storage, as is continually the case in the fuel pools. However, there are likely to be some decommissioning wastes as with other nuclear facilities. Third, since dry-cask storage systems are, for the most part, self-contained, with no mechanical pumps or other active systems, the maintenance of safety relies passively on the cask integrity.

Dry casks do pose their own dangers, as is to be expected. For example, the DOE notes that "rough handling" could result in the release of "small quantities" of gaseous radionuclides from the storage casks.[37] Dry casks also do not eliminate hazards which may result from earthquakes. It is not feasible to expect complete safety when dealing with the extreme hazard represented by the intense radioactivity of irradiated fuel no matter what the storage technology, but using a passive dry storage system is better than having to rely on active mechanical systems that can wear out, malfunction or break down.

Dry storage is currently in use at two nuclear plants under the auspices of a DOE-utility cooperative demonstration program authorized by the Nuclear Waste Policy Act of 1982. The NRC granted licenses in 1986 to the Virginia Power Company for dry storage in metal casks at its Surry nuclear plant near Gravel Neck, Virginia, and to the Carolina Power & Light Company for the use of horizontal concrete modules at its H.B. Robinson plant near Hartsville, South Carolina.

---

35. DOE 1989d.
36. With current U.S. nuclear fuel designs, water is used to moderate the neutrons, thereby allowing the nuclear chain reaction to continue. If there is no neutron moderation, there cannot be an accidental criticality.
37. DOE 1989d, p. I-95.

Both of these plants have loaded irradiated fuel into their dry storage systems. Dry storage in horizontal concrete modules was also licensed at Duke Power's Oconee plant near Seneca, North Carolina in 1990. License applications for spent fuel dry storage facilities are under review for the following five plants as well:[38]

Plant	Utility	Location
Calvert Cliffs	Baltimore Gas & Electric	Annapolis, MD
Brunswick	Public Service of Colorado	Southport, NC
Ft. St. Vrain	Northern States Power	Denver, CO
Prairie Island	Northern States Power	Minneapolis, MN
Palisades	Consumer Power Company	South Haven, MI

There does not seem to be any reason why this form of storage at reactor sites cannot be expanded to accommodate anticipated future spent fuel generation, even in the absence of an MRS or repository in the near term. Indeed, the NRC has recently finalized regulations which essentially encourage such expansion (as long as certain pre-approved cask designs are used) by relaxing previously existing licensing requirements which applied to dry casks.[39] And in considering both the pools and the dry cask option, the NRC has stated that it considers onsite storage to be safe for at least an interim period of 100 years.[40] Of course, careful monitoring should accompany any expansion, since the technology is still relatively new.

It is important to note that dry storage on site does not eliminate the need for spent fuel pools altogether, since recently discharged fuel is so hot, it must be cooled in wet pool storage for one year or more before it can be placed in dry storage.

---

38. NRC 1991, p. 61; and Harvey 1991, p. 16.
39. NRC 1990, pp. 29181-29195.
40. NRC 1989a, p. 39767.

Environmental and Financial Risks of Current Programs

Current repository programs for the permanent disposition of spent fuel, reprocessing wastes, and transuranic wastes encompass two sites—the WIPP site in New Mexico in which the DOE wants to dispose of some of its transuranic wastes, and the Yucca Mountain site in Nevada. Both sites have been confronted with similar scientific, technical, managerial, and environmental questions specifically in regard to DOE management and conflicts of interest. The WIPP site has been built, but has so far not been able to meet requirements of environmental laws and regulations even for temporary experimental waste emplacement. Despite this, the DOE wants to move ahead with the use of this site, and has sought variances from compliance with environmental laws. The Yucca Mountain site has not yet been characterized.

The status of the DOE program at each of these sites is examined in greater detail below.

Transuranic Wastes at WIPP[41]

In 1970, after some 25 years of burying transuranic-contaminated wastes in shallow trenches, the Atomic Energy Commission (the DOE's predecessor agency) decided that transuranic waste was potentially dangerous and therefore unsuitable for shallow-land burial. The AEC began requiring that transuranic waste which contained more than 10 nanocuries per gram be stored in retrievable containers, pending its permanent disposal in a deep geologic repository.

The Waste Isolation Pilot Plant (WIPP) is DOE's geologic repository project for the disposal of transuranic wastes in the salt beds of southeastern New Mexico, at a site about 25 miles from the town of Carlsbad.

The origins of WIPP go back to the early 1970s, when the Atomic Energy Commission moved the focus of its search for waste disposal to New Mexico in the wake of its failure in Lyons, Kansas. Originally conceived as a pilot facility for commercial and military high-level waste disposal, WIPP has since 1979 been slated for use as a disposal

---

41. Much of the material for this section is updated from Saleska 1989, pp. VII-10 to VII-14.

site for military transuranic wastes only.[42] As a DOE project, WIPP is not subject to NRC licensing, but the DOE has agreed with New Mexico that it will be subject to EPA standards.[43] At what point such compliance must be demonstrated, however, has become a point of contention, with the DOE wanting to begin loading waste for "experimental" demonstration purposes before showing compliance. Other federal agencies and independent scientists have questioned the need for this, and assert that compliance with final EPA standards should be shown before any wastes are loaded into WIPP.[44]

Unlike the DOE's program at Yucca Mountain, WIPP is partly built.[45] Located 650 meters below surface, the $1 billion repository consists of a 112-acre underground area, and has a capacity of about 880,000 55-gallon drums, enough to contain slightly less than 160,000 cubic meters (5.6 million cubic feet) of waste.[46]

Numerous technical issues related to the geology and hydrology of the WIPP site and the nature of the transuranic waste intended to be placed there raise questions about its suitability and the DOE's management of the program. These issues include:

------

42. Carter 1987, pp. 177-182.
43. First Modification to the July 1, 1981 "Agreement for Consultation and Cooperation" on WIPP by the State of New Mexico and the U.S. Department of Energy, November 30, 1984. (Reference courtesy Don Hancock, Southwest Research and Information Center, Albuquerque, New Mexico.)
44. GAO 1988b, pp. 10-14, summarizes criticism along these lines from GAO, the National Academy of Sciences, the state of New Mexico's Environmental Evaluation Group, and the Scientists Review Panel, a group of New Mexico scientists.
45. Only about 15 percent of WIPP has actually been mined. This is because the natural phenomenon of "salt creep" (which is the tendency of salt to gradually "flow" and fill empty spaces) causes any rooms mined to close as the salt creeps in to refill the mined space. This gradual room closure is an anticipated part of any waste-disposal process in salt, but the rooms cannot be mined too far in advance of waste emplacement. The DOE therefore plans to mine the additional waste-disposal rooms as the time of permanent waste emplacement approaches. (DOE 1990d, p. 79)
46. The actual capacity of the repository has been the subject of a small controversy. At one point the DOE claimed (DOE 1989c) that the capacity was 1.1 million 55-gallon drums containing about 6.5 million cubic feet. The New Mexico Environmental Evaluation Group, however, estimated that the space only allowed for about 850,000 drums, after which the DOE apparently adjusted its estimate to 880,000 drums containing 5,598,000 million cubic feet (about 158,500 cubic meters). (DOE 1990f, Comments and response section of Vol. 3, and Table 3.1 in Vol. 1)

- *Pressurized water pockets below WIPP.* WIPP is mined out of the lower part of a 2,000-foot thick geological formation called the Salado formation. The Salado is directly above another formation called the Castile formation which has been discovered to contain brine (salt water) pockets under pressure. Boreholes drilled from ground level which have breached these brine reservoirs have typically experienced brine flows to the surface of several hundred gallons a minute.[47]

  The initial WIPP site was abandoned in 1975 when the first WIPP borehole encountered brine and was moved to a different location a few miles away. The planned orientation of WIPP then had to be adjusted in 1981 when the new location was discovered to be within 500 vertical feet of a brine reservoir estimated to contain 700 million gallons of pressurized brine. A 1987 DOE study shows that a pressurized brine reservoir may be present 800 feet below the current repository location, posing the risk that at some point in the future, brine could breach the repository and carry radioactivity to the surface.[48]

- *Water leakage into WIPP.* The DOE first encountered water seepage into WIPP excavations in 1983. In 1986, a member of the National Academy of Sciences (NAS) panel on WIPP warned that in a few hundred years sufficient brine might seep into the repository rooms to saturate them. The water leakage issue became public in the fall of 1987 when a group of New Mexico scientists (called the Scientists Review Panel) concluded that the salt formation at WIPP contains much more water than the DOE had anticipated. They warned that over a period of time the brine could corrode the waste drums, forming a "radioactive waste slurry" consisting of a mixture of brine and nuclear waste which might eventually reach the surface.[49]

---

47. Testimony of Lokesh Chaturvedi, Deputy Director, New Mexico Environmental Evaluation Group, as contained in House 1988.
48. *Ibid.*
49. GAO 1988b, pp. 8-9.

- *Wall cracking and ceiling collapse.* Cracks have appeared in the ceilings and floors of several large waste storage rooms, and in three areas, the ceiling has also collapsed. The cracking and collapse are the result of a rate of room closure two to three times faster than was anticipated. When the first storage rooms were excavated in 1983, the DOE expected it would take 25 years for the creeping salt walls to completely close in on each other, locking the barrels of waste into a mass of solid salt rock. However, at the rate the rooms are closing, it may take them only 13 years or less before complete closure. This rapid rate of closure resulted in the initial cracking, which, although known about by the DOE since 1987, was not publicly revealed until the cracks were also discovered by the New Mexico state Environmental Evaluation Group in May 1989—a month after the DOE closed the rooms to workers because of fears that sections of the ceiling might fail and collapse.[50]

On June 19, 1990, a 100-ton section of the ceiling of a test room did collapse, just 18 days after a technician was in the room for an inspection.[51] A 1,400-ton rockfall occurred in the same area in February 1991, despite the fact that the DOE had installed bolts in the ceiling to improve its stability. As of June 1991, two additional rockfalls have also been reported.[52]

The DOE maintains that the cracks were not important enough to mention publicly because simple mining techniques could make them harmless, and that the ceiling collapses are an expected result of an experiment to see how fast the closure took. But the more significant issue is the existence of yet another unanticipated behavior, which again raises the question: if the DOE is unable to anticipate repository behavior on the scale of a few years, what confidence can there be in its behavior over the next several

---

50.   Schneider 1989, p. 8.
51.   Goessl 1990.
52.   GAO 1991, pp. 10-11.

thousand centuries?

- *Natural resource issues.* Natural resources, especially oil
  and gas, are known to exist in the region of the WIPP
  repository. There are proven reserves in the area, and even
  an existing oil and gas lease beneath the WIPP site.[53] These
  resources could invite future intrusion into the site, a matter
  of particular concern in light of the existence of the pockets
  of pressurized brine discussed above.

The nature and characteristics of the transuranic waste intended to
be emplaced in WIPP also pose potential problems.

- *Compliance with hazardous waste law.* Approximately 60
  percent of the transuranic wastes slated for disposal in WIPP
  are also contaminated with hazardous chemicals.[54] Thus,
  these transuranic wastes are also mixed waste, meaning they
  are also regulated under the law on hazardous wastes, the
  Resource Conservation and Recovery Act (RCRA).[55]

  RCRA regulations require that the disposal of hazardous
  wastes be characterized in detail so that human health and
  the environment are adequately protected. However, according to the DOE, "a large volume of transuranic-mixed wastes to be sent to WIPP were generated in the distant past.
  Documentation on the chemical constituents of these wastes
  ... is often inadequate or does not exist."[56] RCRA, however,
  contains a "land ban" provision which prohibits the disposal
  of uncharacterized or untreated hazardous wastes via land
  burial unless it can be shown that there will be no migration
  of the hazardous materials. Instead of going through the

---

53. House 1991. A recent House bill, H.R. 2637 (passed by the House Interior Com-
    mittee on June 26, 1991) contains a provision (section 11) which would condemn
    the existing lease. The DOE opposed this provision.
54. DOE 1989c, Vol. 2, p. B-22.
55. Regulations pursuant to RCRA are incorporated into 40 CFR Part 264.
56. DOE, RCRA Compliance at the Department of Energy's Waste Isolation Pilot
    Plant, DOE/WIPP 88-018, p. 12, June 1988, as contained in House 1988, pp. 135-
    155.

process of characterizing and treating the wastes, however, the DOE sought a variance from EPA's regulations through a "no-migration" petition. Although not granting the DOE's request for a complete and permanent variance from RCRA's land ban, in the fall of 1990, EPA granted a temporary conditional 10-year variance for DOE's planned experimental demonstration phase at WIPP.[57]

• *Gas generation from waste.* The chemical components of the wastes to be emplaced in WIPP are such that they are continuously, as a result of chemical breakdown processes, generating gases which are released from the waste. If the gas generation rate is sufficient, the concern is that after WIPP has been sealed, the repository could become pressurized, thereby either preventing proper closure from salt creep, or providing a mechanism to forcibly push the waste from the repository. Either way, proper isolation of the waste may not be achieved.[58]

There are serious questions about whether WIPP can meet the EPA's waste disposal standards,[59] particularly due to concerns about the probability of human intrusion some time in the future due to the natural resources in the region, and due to concerns about gas build-up contributing to radionuclide release.[60]

In addition to the technical problems with the site itself, the risks and uncertainties associated with the program have only been exacerbated by the DOE's management. Problems include:

• *Missing documentation.* One part of the process of certifying the WIPP is the completion by the DOE of a safety analysis

---

57.   EPA 1990a, p. 47700.
58.   Telephone conversation between Anthony Gallegos, Performance Assessment Specialist, New Mexico Environmental Evaluation Group, and Scott Saleska, IEER (6 November 1991). Sandia 1990, pp.VI-19 to VI-33, discusses attempts to model gas generation to determine whether it will interfere with the repository's ability to comply with environmental standards.
59.   Those at 40 CFR 191.
60.   Detailed consideration of WIPP's ability to comply with EPA standards is contained in Sandia 1990.

report. However, an internal DOE review of a 1988 draft report found that it "do[es] not contain sufficient information for us to independently conclude that the facility can be operated safely."[61]

Another preliminary review of the available information on WIPP by the DOE's Brookhaven National Laboratories criticized the lack of documentation of a design change that was made in the early stages of WIPP construction by a WIPP project panel. Said the Brookhaven reviewers, "No documentation of this process was found, nor was the listing of panel members made available to us."[62] The Brookhaven review found that no conclusion could be drawn about the safety of WIPP design "due to the lack of documentation available..."[63]

The DOE officially completed what it referred to as a Final Safety Analysis Report in June of 1990. It is not actually final in any meaningful sense, however; even the document itself refers to upcoming amendments and additions that will be required over the next several years as various tests are completed.[64] Outside reviewers have criticized the document as being incomplete even regarding the tests which have been conducted.[65]

- *Construction Quality.* Based on the limited information that was available, the DOE's review found reason to suspect that the construction quality at WIPP may not be adequate. Citing a 1986 accident at WIPP that involved a fire protection system which caused over $100,000 damage, the review noted that "[t]he accident was caused by failure to properly install the fire water system pipes. This occurrence raises questions about general construction quality."[66]

61. Memorandum from James P. Knight, DOE Office of Safety Appraisals (Sept. 2, 1988), as contained in House 1988, p. 116.
62. Brookhaven National Laboratory, Report on Trip to Waste Isolation Pilot Plant, p. 3 (August 24, 1988) (as contained in House 1988, pp. 120-128).
63. Ibid., p. 4.
64. DOE 1990g.
65. Letter from Robert Neill, Director, New Mexico Environmental Evaluation Group, to Arlen Hunt, WIPP Project Manager, August 9, 1991.
66. DOE, Memorandum to James P. Knight through Edward F. Branagan, Jr., regarding "Site Visit to Albuqurque Operations Office and the Waste Isolation Pilot Plant," p. 3, September 1,1988 (as contained in House 1988, pp. 102-108).

- *Misplaced Priorities.* In hearings before the House Environment, Energy, and Natural Resources Subcommittee of the Committee on Government Operations in the fall of 1988, the DOE was criticized for devoting inadequate staff to assess the safety of WIPP and its ability to comply with environmental standards. "There are more DOE people lobbying for legislative authority to emplace waste than there are trying to make sure that facility can be run safely," noted Subcommittee Chairman Mike Synar (D-OK). "That is a formula for disaster."[67]

There are other problems related to the waste capacity of WIPP. The expected capacity of the facility is a little under 160,000 cubic meters.[68] The amount of transuranic waste in retrievable storage at the beginning of 1990 was about 62,000 cubic meters, and the net accumulation of this retrievable waste by the end of the next 20 years or so is expected to be almost 112,000 cubic meters.[69] WIPP's projected capacity is sufficient to accommodate this amount. However, as can be seen from Table 1, page 20 of the chapter on radioactive waste characteristics, this does not include the 190,000 cubic meters of buried transuranic wastes, or the 390,000 to 540,000 cubic meters of transuranic contaminated soil also present at various sites. Thus, the total amount of transuranic waste in all forms expected to be present by early next century is in the range of 443,000 to 592,000 cubic meters—roughly two-and-one-half to three-and-one-half times the capacity of WIPP. The DOE has so far failed to make a determination about what it plans to do with this buried transuranic waste and transuranic-contaminated soil, yet it is this waste—not the retrievably stored waste in monitored facilities—which poses the greatest environmental risk.

Thus, the DOE's current transuranic waste policy is something of a paradox. On the one hand, DOE policy is that transuranic waste is so dangerous that it needs to be stored in a repository 650 meters underground to isolate it from the environment. On the other hand, DOE

---

67.   Exchange between Jill Lytle, Deputy Assistant Secretary for Nuclear Materials, U.S. Department of Energy, and Congressman Mike Synar, House 1988, p. 131.
68.   DOE 1990f.
69.   DOE 1990d, p. 81. The DOE's projections for transuranic waste generation go until the year 2013.

policy so far leaves unaddressed the permanent disposition of the majority of the transuranic waste contaminating the ground or lying in shallow pits and disposal cribs.[70] Because of its planned size, WIPP cannot provide a complete solution for the disposal of many of the transuranic wastes that are causing the greatest contamination problems.

Yucca Mountain

Yucca Mountain is located in southern Nevada about 100 miles northwest of Las Vegas. It is on the edge of the DOE's nuclear weapons testing site, where about 700 announced underground nuclear test explosions have taken place to date.

The geologic formations beneath Yucca Mountain are made of a material called "volcanic tuff," formed from volcanic eruptions occurring between eight and 16 million years ago. The water table is deep, lying as much as 760 meters (2,500 feet) below the surface. Because the water table is so deep, the repository can be constructed above it, in the "unsaturated zone." The DOE believes that the low rainfall (about six inches per year) and high evaporation rate mean that there would be little percolation of water downward through the rocks to the water table.[71]

The Yucca Mountain site is slated for a multi-billion dollar site characterization process to collect data and examine whether it is suitable for a repository. After this process, if the DOE believes that the site is suitable, a license will be submitted to the NRC for construction of a repository.

However, a number of factors relating to both technical and managerial aspects of the Yucca Mountain repository program point to problems inherent to the site, as well as to the way the overall program is proceeding.

Three major technical issues raising serious questions about site suitability are geological complexity, the hydrology of the site (i.e. the behavior of water), and mineral resource concerns.

On the subject of geological complexity, there are at least two prominent concerns:

---

70. This same issue was first raised by the U.S. General Accounting Office in 1986. See GAO 1986c.
71. DOE 1988a, p. 15.

- There are 32 known active faults at the site, including the Ghost Dance Fault, which intersects the proposed location of the underground waste disposal rooms.

- There is potential for volcanic and associated tectonic activity to affect the repository. Several volcanic cones exist near the site. One of them, the Lathrop Wells cone, was believed by DOE geologists to have been inactive for hundreds of thousands of years, until recent research revealed data indicating that it might have been active as recently as 5,000 to 10,000 years ago.[72] This is very recent in terms of geologic time, and certainly within the timeframe during which radioactive waste will pose significant hazards.

  An NRC memo noted that these data "strongly suggest that the various probabilities (and consequences) which have been used for volcanic disruption of a repository at Yucca Mountain may be in error by several orders of magnitude."[73] This is because recent volcanic activity means that future behavior is less predictable, and that there is a greater possibility of a future eruption which would threaten repository integrity.

The hydrology of the site is also an important issue, since the water is a principal potential pathway for the escape of radionuclides. There are at least three aspects to this:

- The rate of rainwater percolation is important. The DOE claims that most of the rainfall evaporates, and the little that remains will take thousands of years to seep through the pores in the rock to the depths of the repository and the water table. Scientists working for the state of Nevada, however, are concerned that the water may actually run along fracture lines in the rock, thereby traveling much more quickly.[74]

---

72.   As reported by Wald 1989, p. A1.
73.   Trapp.
74.   Nevada 1989.

- There are also unresolved concerns about the potential for the water table itself to rise and flood the repository. This issue was first raised by one of DOE's geologists in an internal memo in 1987—before the Nuclear Waste Policy Amendments Act singled out Yucca Mountain as the only site for characterization.[75] Since it became public after that time, the issue has become the focus of much controversy, and a peer review panel is expected to come out with a report on the subject shortly.

- The above concerns about water behavior are further compounded by findings by the 1983 National Academy on nuclear waste isolation. The Academy study concluded that if water at Yucca Mountain does get into the repository site, there is a "major chemical disadvantage" in comparison to other potential sites because radionuclide "solubilities are higher than in other candidate host environments."[76]

An issue affecting hydrological issues in particular is the prospect of unpredictable but significant climate change. Such change, due either to human activities which contribute to the greenhouse effect or to natural climatic variation over the very long hazard-life of the waste, could significantly alter the characteristics of the ground or surface waters.

A third area of concern is mineral resources in the region. This is because potential mineral, oil, and gas resources near the site could invite future human intrusion affecting repository waste isolation.

Additional technical problems concern interactions between aspects of the Yucca Mountain site and the characteristics of the waste forms which would be emplaced there. For example, there may be problems with the disposal of vitrified high-level wastes (glass) in Yucca Mountain, and there is the possibility that carbon-14 releases from spent fuel would violate environmental standards.

A problem with vitrified glass wastes (the disposal form planned for high-level reprocessing wastes) may arise under conditions of slowly increasing humidity in the unsaturated environment at Yucca Mountain. This could provide just the right conditions for chemical decomposition of the surface of the radioactive glass waste form. Sub-

---

75.   Szymanski 1987.
76.   NAS 1983, pp. 185-186.

sequent saturation of the repository (due to a rise in the water table as a result of climate change, for example), could result in rapid transport of radionuclides from the decomposed waste to the accessible environment, allowing the delivery of high doses from groundwater to individuals.[77]

Recently there has been growing concern about the ability of a repository in the Yucca Mountain environment to meet the release limits for carbon-14.[78] For example, a scientist from Lawrence Livermore has written that:

> [I]t is unlikely that waste containers of reasonable cost could meet either the DOE interpretation of 'substantially complete containment ...,' or the NRC 10 CFR 60 release rate limit [of 1 part in 100,000 per year] for carbon 14...
>
> [I]t is also not possible to give assurance that EPA 40 CFR 191 cumulative release limit of 0.1 curie [of carbon-14 per metric ton] over 10,000 years could be met.[79]

This is portrayed, not as a problem for the repository, but as an example of why the current standards are too strict. The author argues that even releases of carbon-14 from the repository far in excess of the current standards would result in doses that are small in comparison to the natural background dose rate which will occur in any case. Therefore, the author writes, "it seems reasonable to conclude that the high level nuclear waste regulations should be changed to allow higher releases of carbon-14 from the repository."[80]

The issue is discussed as if it were a generic problem with the high-level waste standards in general. However, it is actually a problem particular to a repository in the unsaturated zone, and in particular to one in a fractured rock zone like that at Yucca Mountain. This is because carbon-14 can most readily be transported when it is exposed to gaseous oxygen, which readily oxidizes it to radioactive carbon dioxide ($^{14}CO_2$), which is easily transported in the gaseous phase, especially through fractures in the rock. Such conditions would be unlikely in a saturated zone repository, where any carbon-14 escaping from the waste packaging would dissolve in the groundwater, where its release would be limited by the much slower rate of groundwater transport.

77. Makhijani 1990.
78. ADL 1990; DOE 1990e.
79. Van Konynenburg 1991, p. 316.
80. Van Konynenburg 1991, p. 317.

Because the release of gaseous nuclides is likely to be a broader problem for a Yucca Mountain repository, we may expect to see an increasing amount of pressure on regulatory agencies from the DOE and nuclear industry sources to relax the release limits contained in environmental standards, not only for carbon-14, but for other gaseous radionuclides as well, such as iodine-129.

This growing pressure to relax the standards to accommodate the particular nature of the Yucca Mountain site is instructive, especially in light of ongoing DOE assurances that actual construction of a repository at Yucca Mountain is not foreordained, and that it is just a candidate site that will be selected only if it is suitable. A consistent criticism of the 1987 law which targeted Yucca Mountain as the only site for characterization has been that once billions are invested in the site, the institutional momentum to build a repository there will be enormous, regardless of the outcome.

Apart from the inherent technical difficulties associated with the site, the DOE's management problems are also causing difficulties for the project. DOE management has been so far from exemplary that even the Edison Electric Institute which represents nuclear utilities in Washington, D.C. said in 1989 that "we are very concerned that the DOE has spent $2 billion of our money already and just seems to be getting more and more behind."[81]

Recently the NAS Board on Radioactive Waste Management criticized the whole DOE approach at Yucca Mountain, as well as NRC regulations governing characterization and licensing, as overly rigid and inflexible. Although the Board endorsed the concept of deep geologic disposal as "the best option for disposing of high-level radioactive waste," it said that the U.S. was taking an approach that "is poorly matched to the technical task at hand."[82] The Board said that the program as conceived and implemented would be unable to respond to the new information or surprises that were bound to arise from any new technical undertaking of such magnitude. The NAS Board stated that it "is particularly concerned that geological models, and indeed scientific knowledge generally, have been inappropriately applied.... In the face of public concerns about safety... geological models are being asked to predict the detailed structure and behavior of sites over thousands of years. The Board believes that this is scientifically unsound and will

---

81. As quoted in Warren 1989, p. 32.
82. NAS 1990, p. vii.

lead to bad engineering practice." As the NAS Board wrote:

> The U.S. program is unique among those of all nations in its rigid schedule, in its insistence on defining in advance the technical requirements for every part of the multibarrier system, and in its major emphasis on the geological component of the barrier as detailed in 10 CFR 60. In this sense the government's HLW program and its regulation may be a 'scientific trap' for DOE and the U.S. public alike, encouraging the public to expect absolute certainty about the safety of the repository for 10,000 years and encouraging DOE program managers to pretend that they can provide it.[83]

Much of the Board's criticism was directed at the current program's inability to deal with or even properly recognize the technical uncertainties inherent in building a repository. The Board said that the DOE's presumption behind the use of huge databases and computer simulation models seemed erroneously to be that more information and detail would lead to decreased uncertainty. The Board pointed out that this presumption is in contradiction to experience with how scientific and technical knowledge in general advances:

> The studies done over the past two decades have led to the realization that the phenomena are more complicated than had been thought. Rather than decreasing our uncertainty, this line of research has increased the number of ways in which we know we are uncertain. This does not mean that science has failed: we have learned a great deal about these phenomena. But it is a commonplace of human experience that increased knowledge can lead to greater humility about one's ability to fully understand the phenomena involved.[84]

The NAS Board calls for "major changes in the way Congress, the regulatory agencies, and [the] DOE conduct their business," suggesting an alternative approach that recognized the uncertainties in repository development, and was designed for dealing with them, rather than pretending they can be removed, thereby increasing the likelihood

---

83. NAS 1990, p. 1.
84. NAS 1990, p. 4.

of failure when they are encountered.

Risks of Continued Reliance on the DOE

The U.S. Government's problems with the high-level waste repository at Yucca Mountain and the WIPP transuranic waste repository near Carlsbad are just the latest in a series of troubles in its attempts to site and build a long-lived waste repository. Taking a broader view of the U.S. program, there is a pattern of consistent slippage and failure.

Table 7, for example, shows how U.S. time tables for the opening of a high-level waste repository have repeatedly slipped. In fact, as the table indicates, the date of projected repository availability seems to be receding further into the future the more time passes. Between 1975 and 1989, while 14 years passed, the repository went from 10 years to 21 years into the future.

Shortly after the most recent repository delay, DOE Secretary James Watkins said, in explaining the delay, that "the whole set of schedules was not scientifically sound, not fiscally sound, not technically sound... They were incomplete, misleading, and not properly done."[85]

Similar problems confront disposal of transuranic wastes at the WIPP site in New Mexico. Though this has been built in the face of numerous objections, its opening date has repeatedly been delayed in the face of failures of the site to comply with environmental laws and regulations, and even the DOE's own procedures. Once again, however, the DOE appears to be giving priority to weapons production activities and trying to override legal and environmental concerns. In the face of the difficulties posed by EPA's hazardous waste regulations,[86] for example, the DOE has sought and received a partial variance from these regulations, as we discussed in the section on WIPP.

As timetables have slipped, costs for the waste disposal programs have increased greatly. Table 8, for example, shows the DOE's estimates for the high-level waste life-cycle system costs, and how they have grown over the years. From an average of $24 billion in 1983, expected 1988 constant-dollar costs for a two-repository system have grown by over 40 percent to almost $34 billion in 1990. Cost escalation would be much worse than this had not significant program chan-

85. As quoted in Wald 1989, p. 8.
86. Promulgated under the Resource Conservation and Recovery Act (RCRA).

Table 7		
RECENT HISTORY OF NUCLEAR WASTE REPOSITORY TARGET DATES		
Year of Estimate	Estimated Repository Availability	Difference
1975-1977	1985	8-10
1980	early 1990's	10+
1982	1998	16
1988	2003	15
1989	2010	21

Sources: The 1985 target date was established by ERDA in 1975 [Lipshutz 1980, p. 140], and was still part of an October 1977 Department of Energy announcement of its new spent fuel policy, whose major thrust was President Jimmy Carter's deferral of commercial reprocessing [Carter 1987, pp. 133-134.]; the early 1990's target date was part of the report of the Interagency Review Group on Nuclear Waste Management [Interagency Review Group, *Report to the President*, TID-2944Z (March 1979), US Department of Energy], which was essentially endorsed by President Jimmy Carter's Nuclear Waste Policy Statement of February 12, 1980 [Carter 1987, pp. 135-143]; the target date of 1998 for a first repository was set in the Nuclear Waste Policy Act of 1982, Public Law 97-425; the 1998 target date was slipped to 2003 by the DOE in 1988, as stated in US Department of Energy, *Draft 1988 Mission Plan Amendment*, DOE/RW-0187, p. 1 (June 1988); the 2010 target date was set in late 1989 by the DOE in US Department of Energy, *Report to Congress on Reassessment of the Civilian Radioactive Waste Management Program*, DOE/RW-0247, p. vii (November 1989).

ges occurred since the DOE made its first comprehensive cost estimate. The most significant of these changes include a reduction in the amount of waste expected to be generated during the life of the program (from 134,000 metric tons to about 86,800 metric tons of commercial spent fuel), and the 1987 Congressional designation of Yucca Mountain, Nevada as the sole site to be characterized for the first repository. This meant that the cost of characterization (which originally included plans to characterize three candidate sites for the first repository) were much reduced from what they would have been had the original assumptions remained. This, however, has had a steep price: increased risk of failure and increased environmental risk. Thus, the program is now on a course of both higher costs and higher financial and environmental risks.

When basic assumptions about waste disposal are kept reasonably constant, the actual cost escalation is probably at least double what it appears to be in Table 8. Table 9 compares the unit costs under the 1983 assumptions to the projected unit costs of the DOE's latest estimate. It can be seen that real costs for the basic two-repository system grew by over 80 percent in eight years on a "per unit of fuel disposed" basis, from $179,100 per metric ton in 1983, to $325,200 per metric ton today.

Cost escalations have also plagued the WIPP facility, and are con-

Table 8
HISTORY OF DOE LIFE-CYCLE COST ESTIMATES FOR A
TWO-REPOSITORY SYSTEM
(Estimates in billions of 1988 dollars)

Year of Estimate	Development & Evaluation	Transportation	Repository Construction	MRS	Benefits Payments	TOTAL
	Major Cost Category					
1983	5.8	4.8	13.1-13.7	NA[a]	NA[b]	23.7-24.3
1984	9.0	3.0-4.6	12.4-15.2	NA[a]	NA[b]	24.7-28.8
1985	8.9	3.8-5.8	14.2-19.2	NA[a]	NA[b]	27.1-33.8
1986	10.1-10.4	1.9-2.5	13.1-21.7	3.1-3.2	NA[b]	28.9-37.4
1987	15.8-16.1	2.1-2.4	13.5-20.1	2.9	NA[b]	34.5-41.0
1989	13.1	2.3	13.4	2.3	0.9	32.0
1990	15.0	2.7	13.6	1.6	0.8	33.6

Source: GAO 1990, p. 19; DOE 1990a, p. 4.
*Based on a no-new-orders scenario with two repositories.
(a) A DOE estimate which included an integral MRS system was not made until 1986.
(b) Benefit payments were authorized by the Amendments Act of 1987.

tinuing at an alarming rate. In just the past two years, for example, the DOE's projection for WIPP expenditures for the five-year period 1991 to 1995 increased by a factor of 67 percent, from $531 million in 1989, to $884 million in 1991.[87] This increase has occurred in spite of the fact that WIPP has not opened as the 1989 cost estimates anticipated.

A more accurate gauge of the true cost escalation may be a comparison of the five-year projection made in 1989 (for fiscal years 1991 to 1995) with the five-year projection made in 1991 (for fiscal years 1993 to 1997), since both estimates anticipate the imminent opening of WIPP. If this comparison is made, five-year cost projections have risen from $531 million in 1989 to around $1.1 billion in 1991, a 107 percent increase.[88] Thus, in just two years, the DOE's cost estimates for the first several years of WIPP operation have more than doubled.

---

87. DOE 1989a; DOE 1991b.

88. DOE 1989a; DOE 1991b. The later estimate is given in a range of $1,048 million to $1,143 million.

Table 9
UNIT DISPOSAL COST ESTIMATES FOR SPENT FUEL
AND HIGH-LEVEL WASTE

1990 Estimates		1983 Estimate
No-New-Orders (96,300 MTU)*:		(134,000 MTU)*
1 repository	$265,800/MTU	
2 repositories	$348,900/MTU	$179,100/MTU**
Upper-Reference (106,400 MTU)*		
2 repositories	$325,200/MTU**	

Source: DOE, as cited in GAO 1990.
* The metric ton equivilence figures cited above include both commercial spent fuel and about 9,500 metric tons of military and commercial spent fuel that has been reprocessed and is expected to be in the form of glass.
**These two repository scenarios probably represent the most similar for purposes of comparing the 1983 and 1990 cost estimates.

Such cost escalations are typical of the DOE's past performance with new programs. It is quite possible that real costs could go on rising significantly, if this past performance is any guide.

B. Low-Level Waste

History of Low-Level Radioactive Waste Disposal

Most low-level radioactive waste generated in the U.S. during the past 40 years has been disposed of by shallow land burial,[89] in which wastes are stored in drums or other containers and typically buried in trenches at depths ranging from 3 to 40 feet. So far, three of the six commercial disposal sites in the U.S. have been closed, and are experiencing environmental problems. At each of the three sites (located at West Valley, New York; Maxey Flats, Kentucky; and Sheffield, Illinois), water has leaked into the burial trenches and in some cases caused extensive movement of radionuclides into the surrounding environment. Rather than being maintenance-free stabilized landfills, as was intended, these sites have ended up requiring active maintenance and remedial activities within ten years of closure.[90]

The problems at Maxey Flats, which was first opened in 1962, provide an instructive example. A 1974 report by the state of Kentucky found that radioactive materials, including plutonium, had moved hundreds of feet from where they had been buried. Although the operator of the site, U.S. Ecology (formerly the Nuclear Engineering Company, or NECO), had claimed that significant subsurface migration of plutonium was not possible, a 1975 report by the EPA found plutonium in core drilling samples, monitoring wells, and drainage streams. The EPA report noted that although Maxey Flats had been "expected to retain the buried plutonium for its hazardous lifetime," the plutonium had actually migrated from the site "in less than ten years."[91]

The state finally closed Maxey Flats in 1977, and the site, which has since been placed on the Superfund National Priorities List by the EPA, is currently undergoing an expensive remediation program. In addition to the $15 million already spent on remediation activities at the

---

89. Although the DOE estimates that some 90,000 containers of low-level radioactive waste were dumped at sea in the 1950s and 1960s. (DOE 1990d, p. 109.)

90. For more details on problems at existing sites, see Resnikoff 1987, Chapter 2, pp. 33-44.

91. U.S. Environmental Protection Agency, Preliminary Data on the Occurence of Transuranium Nuclides in the Environment at the Waste Burial Site, Maxey Flats, Kentucky, EPA-520/3-74-021, Washington, DC: EPA Office of Radiation Programs, February 1975, as cited in Resnikoff 1987, p. 35; and Lipshutz 1980, p. 132.

site, official estimates of what the total remediation effort will require range from \$34 million to \$70 million in 1989 dollars (discounted at an annual rate of 4 percent).[92] When the clean-up is finally done and all the costs accounted for, final disposal costs for the wastes at Maxey Flats may well be roughly 10 to 50 times greater than the original fee charged to bury them there.[93]

Current Standards and Regulations

In the wake of the problems experienced at low-level waste burial grounds, new NRC regulations for the land disposal of low-level radioactive waste were developed, and were issued in 1983.[94] According to the NRC, the reason for the new regulations was that the previously existing ones "[did] not contain sufficient technical standards or criteria for the disposal of the licensed materials as waste."[95]

However, although the new standards may represent an improvement over the practically non-existent ones of the 1960s and 1970s, they are still fundamentally flawed. To understand these problems, we first must review the waste classification and disposal standards as they now exist.

The ABCs of Low-Level Waste Classification

As mentioned previously, low-level waste is a catch-all category of radioactive waste that is not actually defined with any reference to its

---

92. Maxey Flats Steering Committee, Feasibility Study Report, Table 4-5, April 1991 (obtained courtesy Marvin Resnikoff, Radioactive Waste Management Associates, New York, New York). Current expenditure estimate of \$15 million provided by personal communication from Marvin Resnikoff (May 1991).
93. According to DOE 1990d, p. 114, about 4.78 million cubic feet of LLW have been disposed of at Maxey Flats. Total clean-up costs cited in text are \$50 to \$85 million (adding the \$35 to \$70 million discounted costs directly to the \$15 million already spent). Resnikoff 1987, p. 36, cites an estimate of \$121 million. This gives net disposal costs ranging from \$10 to \$25 per cubic foot of waste disposed. Disposal costs in 1975 were \$1 per cubic foot (OTA 1989). Presumably, disposal costs were significantly lower when Maxey Flats started operation in 1962; we assume \$0.50 to \$1 per cubic foot which results in a disposal cost escalation factor ranging from 10 to 50.
94. NRC 1983a.
95. As cited in OTA 1989, p. 59.

"level" of radioactivity, but instead includes any waste that does not fall into other categories.

However, for the purposes of management and disposal of low-level waste, federal regulations do divide it into four classes which are determined by radioactivity level and longevity of half-life. These classes, as previously mentioned, are, in order of increasing hazard, named Class A, Class B, Class C, and Greater-than-class-C.

Table 10 contains the NRC radionuclide concentration limits which define the various classes of commercial low-level waste. Waste which only contains radionuclides in concentrations below their Class A limits is Class A waste. Low-level waste containing any radionuclide whose concentration exceeds the Class C limits for that nuclide is Greater-than-class-C waste.

Table 10
NRC LIMITS DEFINING CLASS A, B, AND C LLW
(Curies per Cubic Meter)

A. "Long-lived Radionuclides"	Half-life (years)	Class A	Class B	Class C
Carbon-14	5,700	0.8	N/A	8.0
Carbon-14 in activated metal	5,700	8.0	N/A	80.0
Nickel-59 in activated metal	75,000	22.0	N/A	220.0
Niobium-94 in activated metal	30,300	0.02	N/A	0.2
Technetium-99	213,000	0.3	N/A	3.0
Iodine-129	15.7 million	0.008	N/A	0.08
Alpha-emitting transuranics with half-lives greater than 5 years		10.0*	N/A	100*
Plutonium-241	14	350.0*	N/A	3,500*
Curium-242	163 days	2,000*	N/A	20,000*
B. "Short-lived Radionuclides"				
Tritium	12.3	40	no limit	no limit
Cobalt-60	5.3	700	no limit	no limit
Nickel-63	100.1	3.5	70	700
Nickel-63 in activated metal	100.1	35	700	7,000
Strontium-90	28.5	0.04	150	7,000
Cesium-137	30	1	44	4,600
Total of all nuclides with less than 5-year half-life		700	no limit	no limit

Source: NRC 1988 (10 CFR Part 61.55).
* Units are in nanocuries per gram (note that Pu-241 and Cm-242 have long-lived daughter products).
**There are no limits established for these elements in Class B or C wastes. If waste is contaminated with these radionuclides in concentrations greater than their Class A limits, the waste is Class B, unless the concentrations of other radionuclides determine the waste to be Class C or above independent of these nuclides.

As the table shows, the NRC divides the radionuclide contaminants of concern into what it refers to as "long-lived" and "short-lived" radionuclides.[96] The long-lived limits in the table are determining, unless the "short-lived" radionuclide concentrations would place the waste in a more hazardous category.

For example, if low-level waste contains any "long-lived" radionuclide in concentrations greater than its Class A limits, and since there are no Class B limits defined for the "long-lived" nuclides, it is Class C waste (provided all concentrations are still below their Class C limits). Only if all "long-lived" radionuclide concentrations are below their respective Class A limits may the waste be classified as Class B if its concentrations are less than the "short-lived" radionuclide Class B limits.[97]

In other words, Class B "low-level" waste may contain "short-lived" radionuclides in concentrations up to the Class B limits specified in the lower half of Table 10 . However, Class B low-level waste may not contain long-lived radionuclides in concentrations greater than Class A limits for "long-lived" nuclides in the upper half of this table. If any of the radionuclides in the "long-lived" category are present in concentrations greater than the limits for Class A, the waste is defined as Class C, or "Greater than class C" depending on the concentration in comparison to the Class C limits.

(Note that plutonium-241 and curium-242 decay into long-lived

---

96.   "Short-lived" and "long-lived" are the designations used in the NRC regulations. It should be noted, however, that the "short-lived" category includes nickel-63, which has a half-life of over 100 years, meaning it could present a potential hazard for about 10 times that long, or over 1,000 years.

97.   For wastes containing mixtures of radionuclides, a sum of fractions rule is followed. This means that the sum of all nuclide concentrations, each measured as a fraction of its limit for the class being considered, must be less than one in order for that class to apply. For example, consider a waste contains 100 curies/m^3 of strontium-90, and 22 curies/m^3 of cesium-137. Both of these concentrations exceed the Class A limits, so they must then be compared to the Class B limits (150 curies/m^3 for strontium-90, and 44 curies/m^3 for cesium-137). The strontium-90 fraction is 100/150, or 0.67; the cesium-137 fraction is 22/44, or 0.5. Since the sum of these fractions, 1.17, is greater than one, the waste may not be Class B, even though the individual concentrations are each below their respective Class B limits. Repeating the same process with the Class C limits results in a value of 0.019—less than one—so the waste may be classified as Class C.

radionuclides, which is why they are in the "long-lived" category despite half-lives which are shorter than many elements in the "short-lived" category.)

NRC Classification And Disposal Standards[98]

The NRC regulations contain standards for land disposal of low-level radioactive wastes, and set specific technical requirements for near-surface disposal (less than 30 meters deep) of this waste.[99] These technical requirements vary by waste class, ranging from Class A, which has the least stringent packaging and disposal requirements, to Greater-than-class-C, which is "generally considered unacceptable for near-surface disposal." However, a careful examination of these standards reveals fundamental inconsistencies that raise serious questions about their adequacy.

Regarding the disposal site, the NRC regulations identify two principal methods of control to prevent excessive radiation exposure over the years to "inadvertent intruders" who might "occupy the site in the future and engage in normal pursuits without knowing that they were receiving radiation exposure." These two methods are: 1) "institutional control over the site after operations by the site owner to ensure that no ... improper use of the site occurs...."; or, 2) disposing of waste which would present an "unacceptable risk" to an intruder "in a manner that provides some form of intruder barrier that is intended to prevent contact with the waste."[100]

The NRC regulations incorporate both types of controls. On the one hand, for example, they state that:

> Institutional control of access to the site is required for up
> to 100 years. This permits the disposal of Class A and Class
> B waste without special provisions for intruder protection,
> since these classes of waste contain types and quantities of

98.   Much of this section is an expansion of the criticism raised in Saleska 1989, pp. III-4 - III-5.
99.   NRC 1988b (10 CFR Part 61).
100. NRC 1988b (10 CFR Part 61.7[b][3]).

> radioisotopes that will decay during the 100-year period and
> will present an acceptable hazard to an intruder.[101]

In addition, the regulations state that "[w]aste that will not decay to levels which present an acceptable hazard to an intruder within 100 years is designated as Class C waste."

Class C waste, must, for this reason, be disposed at a greater depth than other classes, or, if that is not possible, under an intruder barrier with an effective life of 500 years. "[A]t the end of the 500 year period," according to the NRC regulations, "remaining radioactivity will be at a level that does not pose an unacceptable hazard to an intruder or public health and safety."[102]

In examining the NRC regulations, one is thus led to believe that the class limits listed in Table 10 were derived from the requirements imposed by these hazard definitions and time frames. However, even according to the NRC's own definitions of what is "hazardous" and what is "acceptable," the time frames of 100 and 500 years are logically incompatible with the class limit definitions, raising serious questions about their environmental and public health adequacy.

For example, as can be seen from Table 10, much of the "100-year" waste (waste Classes A and B), for example, will not decay to NRC-defined "acceptable" levels in 100 years. Consider nickel-63. Buried at Class B concentration levels of just under 70 curies per cubic meter, waste containing nickel-63 would still have a concentration of about 35 curies per cubic meter after the institutional control period of 100 years had elapsed. According to the NRC regulations, at this point the waste should have decayed to the point where it "will present an acceptable hazard to an intruder." Yet, at 35 curies per cubic meter, the waste, if retrieved from the disposal site and re-buried, would still be classified as Class B waste, since it has concentration levels which are 10 times higher than the Class A limits. As a matter of fact, this waste would take a total of well over 400 years to decay just to the Class A upper limits

---

101. NRC 1988b (10 CFR Part 61.7[b][4]). It should be noted that in another part of the regulations, the assurances here about a 100-year limit to the hazard not-withstanding, Class B is lumped with Class C, where the regulations state that for both of these classes, their "waste forms or containers should be designed to be stable, i.e., maintain gross physical properties and identity, over 300 years". [10 CFR 61.7(b)(2)]. The NRC does not explain why waste that is supposedly hazardous for only 100 years is required to be in a form or container that will last 300 years.

102. NRC 1988 (10 CFR Part 61.7[b][5]).

(at which point the NRC regulations would still define it as hazardous for another 100 years if it were being buried for the first time).

This analysis makes an even stronger case against the NRC regulations when applied to the Class C limits in Part A of Table 10, which pertains to "long-lived radionuclides." Class C waste, according to the NRC, is 500-year waste. Consider Class C waste contaminated with technetium-99, however. Buried at concentrations of just under the Class C limit of 3 curies per cubic meter, this waste will be hazardous according to NRC definitions for far longer than 500 years. It will take such waste over three half-lives—some 640,000 years—just to decay to the upper boundary of Class A levels.

The illogical nature of the above regulatory approach is made even more explicit in the NRC's discussion of the "long-lived" radionuclides in the waste. According to the NRC, in managing low-level waste,

> consideration must be given to the concentration of long-lived radionuclides... whose potential hazard will persist long after such precautions as institutional controls, improved waste form, and deeper disposal have ceased to be effective. These precautions delay the time when long-lived radionuclides could cause exposures.[103]

In essence, here is an admission that the hazard due to long-lived radionuclides "will persist long after" the controls imposed by the regulations fade away. This is an extraordinary admission of the regulation's fundamental inadequacy right in the text of the regulation. The only thing the NRC regulations will apparently do with respect to the long-lived components of low-level waste, is push the hazard into the future, since NRC-mandated controls will, at most, only "delay the time when long-lived radionuclides could cause exposure." In the case of many long-lived radionuclides, they will continue to be present in almost exactly the same concentrations when institutional controls have lapsed as when they were first buried.

Clearly such regulations are inconsistent and do not provide a sound scientific basis for addressing the problems of radioactive waste disposal.

The orphaned Greater-than-class-C wastes appear to have no clear plan currently for their disposal. They are the most hazardous class, and

---

103. NRC 1988b (10 CFR Part 61.55[a][1]).

for this reason are nominally slated for repository disposal by law and NRC regulation.[104] However, the DOE does not appear to be actively making plans for accepting them at Yucca Mountain, nor is it including these wastes in its estimates of repository system cost.[105] The DOE's position appears to be that it can exclude "greater than Class C" wastes from its planning for Yucca Mountain because the Nuclear Waste Policy Act does not mandate inclusion of these wastes.

Many of the problems with the current waste classification and disposal regulations in the U.S. derive from one simple factor. The various categories of waste are basically defined according to the process which produced them (uranium milling, reprocessing, etc.) and not according to the longevity or concentration of the radioactive materials. Thus we find ourselves in the rather odd situation of preparing to dispose of very hot, long-lived waste in shallow land burial, where it may eventually cause environmental and health damage, while at the same time consigning comparable or even less radioactive wastes to deep transuranic and high-level waste repositories.

No EPA Standards for Low-level Waste

It is noteworthy that there are no EPA standards which apply to "low-level" waste disposal. In this respect, the regulatory status regarding "low-level" wastes is even worse than the corresponding status regarding high-level and transuranic wastes.

The EPA has authority to develop such standards, and in 1983 (the same year that the NRC low-level waste regulations were promulgated) published its intention to develop generally applicable standards for low-level radioactive waste.[106] The EPA proceeded to develop a draft

---

104. The 1985 Low-level Radioactive Waste Policy Amendments Act (LLRWPAA) assigned responsibility for the disposal of Greater-than-class-C waste to the federal government—that is, to the DOE. NRC regulations "require disposal of greater-than-class-C low-level radioactive waste in a deep geological repository unless disposal elsewhere has been approved," (NRC 1989b, p. 22578).

105. The DOE's cost assessment of the repository program only includes consideration of spent fuel, high-level reprocessing waste, and disposal of low-level waste generated in the process at conventional low-level waste burial sites. The study's assumptions explicitly state that "Other types of potentially high-level wastes have not been included" in the analysis (DOE 1989b, p. B-3).

106. Referred to in EPA 1990b, p. 1.

proposal for standards, but according to EPA officials, the agency has
so far "been unable to issue it for public comment because of continuing unresolved differences with sister federal agencies."[107] This appears to be a reference to NRC and DOE disagreements with the EPA
which has led the federal Office of Management and Budget (OMB) to
prevent them from being officially released. The NRC and DOE objections are apparently due to the fact that the proposed EPA standards are
more stringent than existing NRC and DOE regulations.

Officials from the EPA say that their standard would improve on
the existing regulatory regime for low-level waste in several ways:[108]

- In addition to actual disposal, it would cover pre-disposal
  management, treatment and storage.

- The EPA standard is broader than existing regulations,
  covering DOE military low-level wastes as well as commercial wastes and setting standards for Naturally occurring and
  Accelerator-produced Radioactive Materials (NARM).

  (As previously mentioned in the section on radioactive waste
  characteristics, NARM wastes are orphan wastes not consistently regulated under any current standard, and in some
  cases they fall under no specific regulations at all because
  they occur outside the nuclear fuel cycle. NARM includes
  such materials as radium-226 and thorium-230 produced
  outside the nuclear fuel cycle, and radionuclides produced
  by particle accelerators. The EPA standard would require the
  disposal of high-concentration NARM wastes—i.e. greater
  than 2 nanocuries per gram—in a regulated, licensed disposal facility.)

- The EPA standard would establish a preventive approach to
  groundwater protection. This is an aspect that is entirely ab-
  sent from the NRC standards, and, according to EPA officials, "is a part of [the EPA] draft standard that NRC strongly

---

107. EPA 1990b, p. 7.
108. EPA 1990b.

opposes." The EPA, however, warns that unless a stronger approach is taken towards groundwater protection, "many [low-level waste] disposal sites may end up on future Superfund lists."[109]

Although it is not possible to judge these claims in the absence of the actual draft standard, it is clear that in the absence of more comprehensive standards than currently exist, there are many loopholes and deficiencies that will go unaddressed.

For example, low-level wastes generated at DOE facilities are not subject to the NRC regulations which apply to commercial waste, but are instead governed by internal DOE waste management policies which have historically been even more lax than the commercial NRC standards. The more recent stricter DOE waste management policy (issued in September 1988 but not expected to be fully implemented for several more years) does establish policies roughly parallel to the current NRC standards. Yet even these new DOE standards still allow burial of low-level waste in cardboard boxes under some circumstances, a practice that has been forbidden by NRC standards for commercial waste.[110] The EPA standards could address such inconsistencies by imposing across-the-board regulations that would apply to both commercial NRC licensees and DOE facilities.

Additionally, as pointed out by the EPA, the advent of surcharges and volumetric quotas for low-level waste generators imposed by the 1985 Low-level Radioactive Waste Policy Amendments Act has increased incentives for the creation of specialized away-from-generator facilities for processing, treatment, and storage of low-level waste. This means that pre-disposal management and storage activities are likely to increase greatly in coming years, resulting in greater potential for off-site exposure, spillage, and exposures from airborne effluents from centralized offsite storage and/or incineration facilities. Existing regulations governing such activities are fragmented. Further, because such facilities are not considered part of the nuclear fuel cycle, they are

---

109. EPA 1990b, pp.4-5.
110. DOE 1988b, p. III-7.

exempt from EPA's 25 millirem annual dose limit which applies to nuclear fuel cycle activities.[111] Instead, such facilities would only be subject to the 100 millirem limit imposed by NRC's standards.[112]

The complete absence of more comprehensive and stringent EPA standards is particularly critical at this time. This is because, as we will discuss below in the section entitled "Current Status and Problems With Disposal Plans," a number of new low-level waste management and disposal facilities are in the planning stages now.. It is at this stage, during the siting, planning and design of a new generation of waste facilities, when comprehensive regulations can most effectively and efficiently have their intended protective effect. Failure to implement such standards now is only another example of how current U.S. waste policy, taken as a whole, is increasing both environmental risks and threatening to result in excessive financial expenditures in the future when we are forced to clean up the consequences of those additional risks.

Current Status and Problems with Disposal Plans

Despite the problems in the U.S. regulations, current law requires the states and regions to move forward with siting and building new disposal facilities.

Regional Compacts for Commercial Waste

Following the closure of three of the six commercial "low-level" radioactive waste dumps in the 1970s, the states where the three still-operating sites were located began to lobby for a greater distribution of the "low-level" waste disposal burden. A number of packaging and transportation incidents involving "low-level" waste in 1979 highlighted the responsibility that these three sites were shouldering for the rest of the country. These incidents caused the governors of Nevada and Washington to close temporarily their states' disposal sites, and the

---

111. In EPA's regulations at 40 CFR Part 190.
112. NRC 1991 (10 CFR Part 20). Note that the 100 millirem dose limit for individual members of the public from NRC licensees is the result of recent changes to the NRC's standards for protection against radiation at 10 CFR Part 20. The old standards which had been in effect from the 1960s, limited doses to members of the public from NRC licensees to 500 millirems.

governor of South Carolina to institute a 50 percent volume reduction at the Barnwell site.[113]

This precipitated a "low-level" waste disposal crisis which forced the issue onto the agenda of the U.S. Congress. The result was the Low-Level Radioactive Waste Policy Act of 1980 (LLRWPA).[114] This act, together with much more detailed amendments to it in 1985,[115] forms the basis for low-level radioactive waste disposal plans in the U.S. today.

These laws established that each state was responsible for assuring adequate disposal capacity for the commercial low-level radioactive waste that was generated within its borders. However, since far fewer than 50 separate disposal sites are needed, the laws also encouraged the formation of voluntary regional state compacts. Each compact could then site one disposal facility in a compact member state for the use of the whole compact.

Thus far, nine compacts involving 42 states have formed. With the exception of the Northwest compact, these are in various stages of siting new disposal facilities.[116] The remaining eight states are either planning to site their own individual disposal facilities, or hope to join an existing compact. The status of compacts and individual states as of late 1990 is indicated in Figure 7.[117] The situation has changed slighty since then. In July of 1991, Michigan was ejected from the Midwest compact, and now is unaffiliated; Ohio has become the designated host state for the Midwest compact.

In order to enforce the siting and construction of new disposal sites, the law contains a number of milestones, along with associated incentives and penalties for those states that meet or fail to meet them.[118] The

---

113. Jordan 1984, p. 7.
114. LLRWPA 1980.
115. LLRWPAA 1986.
116. The Northwest compact, consisting of Washington, Oregon, Montana, Idaho, and Utah, will continue to use the existing Richland commercial disposal facility at Hanford Nuclear Reservation. The other two disposal facilities, at Barnwell, South Carolina and Beatty, Nevada, plan to close at the end of 1992, necessitating the construction of new sites for their compacts.
117. DOE 1990c, p. vi.
118. These incentives and penalties include surcharges and rebates on waste disposed outside the compact, and the threat of a cut-off of access to disposal at existing disposal sites.

Figure 7

LOW-LEVEL WASTE COMPACT STATUS, 1990

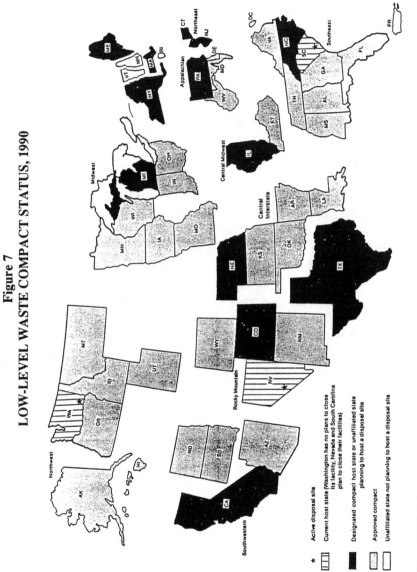

* Active disposal site

▢ Current host state (Washington has no plans to close its facility, Nevada and South Carolina plan to close their facilities)

■ Designated compact host state or unaffiliated state planning to host a disposal site

▨ Approved compact

▢ Unaffiliated state not planning to host a disposal site

Source: DOE 1990c, p. vi.

incentives and penalties become increasingly stronger as each milestone is reached. The bottom-line enforcement mechanism is the requirement that a state assume ownership and possession of all wastes generated within its borders if provisions for the disposal of this waste have not been made by the final deadline of January 1996 (there is an initial deadline of January 1993). Furthermore, according to law, such a state "shall be liable for all damages directly or indirectly incurred by [the waste generators] as a consequence of the failure of the State to take possession of the waste."[119]

DOE Military Wastes

Military low-level wastes routinely generated within the DOE nuclear weapons complex are buried, as mentioned previously, according to the DOE's own internal rules. Although roughly parallel to the NRC's regulations, they are in some respects less stringent.

Although these practices are cause for concern, some of the DOE's future plans for waste disposal are even more troubling, and appear to verge on contradicting the apparent meaning of current law. For example, as we have discussed, federal law clearly defines "high-level radioactive waste" (which is required to be disposed of in a repository) as including "liquid wastes resulting from ... reprocessing irradiated fuel."[120] Nonetheless, the DOE plans to separate some of the radioactivity in Hanford's high-level liquid reprocessing wastes and convert it into a solid grout form. The DOE considers this grout to be "low-level" waste, which they intend to bury in shallow vaults.

The quantities of radionuclides planned to be grouted at Hanford as of early 1989 were enormous, and included:[121]

- 12 to 20 million curies of cesium-137.

- 1 to 8 million curies of strontium-90.

- 30 to 150 kilograms of plutonium.

---

119. LLRWPAA 1986, at Section 5(d)(2)(C).

120. NRC 1988b (10 CFR 60.2).

121. Wodrich 1989, pp. 4,8.

For comparison, the entire grout campaign at Savannah River en-
visages a grout discharge campaign whose accumulation peaks at about
115,000 curies.[122] For further comparison, according to current DOE
records, the cumulative total radioactive inventory of all radionuclides
in all AEC/DOE low-level radioactive waste disposed at all active DOE
disposal sites in the U.S. through 1989 amounts to about 14 million
decay-corrected curies.[123]

Thus, the Hanford grant campaign, if actually undertaken, may bury
twice as much radioactivity as now remains in all major DOE low-level
waste sites from decades of operation. This is a huge amount of radioactivity that by any reasonable standards should be considered repository-
destined long-lived waste. That such a proposal could be seriously
considered shows again the danger in the present lack of standards, and
the need for ones which are clear and identifiable.

Recent experiments with grout are showing the danger of this approach. Organic chemicals which were added to the high-level waste
as part of past waste treatment operations at Hanford are causing
problems in the formation of the grout, which is a cement-like mixture.
Organic chemical breakdown is leading to the evolution of hydrogen
gas from the grout, and Hanford personnel have had to put pipes into
the experimental grout mixtures to vent the hydrogen gas. This indicates there may be a serious potential for rapid cracking and disintegration of the grouted waste form. Further, the experiments show that the
organic chemicals and nitrate compounds in particular are highly sus-
ceptible to leaching out of the grout, posing a groundwater pollution
problem, especially from the nitrates.[124]

---

122. A peak of 116,000 curies is projected to occur in 2006 (DOE 1990d, p. 113).
(Grout at Savannah River is generally referred to as "saltstone.")
123. DOE 1990d, Table 4.1, p. 105. The sites included (as listed on DOE 1990d, p.
107) where low-level waste defense wastes are buried are: Hanford, Savannah
River, Idaho National Engineering Laboratory, Oak Ridge (including sites for the
National Lab, the Gaseous Diffusion Plant, and Y-12 plant), Fernald, Pantex,
Nevada Test Site, Los Alamos, Lawrence Livermore, Paducah, Portsmouth, Sandia National Labs, and Brookhaven National Labs.
124. This issue was discussed in some detail at the meeting of the DOE's Technical
Advisory Panel on Hanford High-Level Waste tank, in Chicago on September 5,
1991. Arjun Makhijani, personal notes.

High Costs, High Risks

The approach embodied by federal law for commercial low-level wastes forces taxpayers at the state level to take on liabilities in an area where corporate and federal governmental failures have led to considerable problems. It has resulted in a politically unstable, costly, and environmentally risky program—especially as far as the longer-lived components of the wastes are concerned.

For example, when Governor James Blanchard of Michigan halted siting activities and threatened to pull out of the compact in early 1989, he alarmed compact members and observers of the compact process across the nation. Michigan's action illustrates the instability of the current situation. The possibility that a state may suddenly pull itself out of a compact continues to pose the threat of a domino effect which could destroy the basis of the low-level waste disposal law. One state low-level waste official, remarking on Michigan Governor Blanchard's threat to pull out of the Midwest compact, said "he was in danger of bringing down a real house of cards."[125]

Although the Michigan situation appeared to be resolved for a time, the state's continued reluctance to move forward on siting a disposal facility caused the state to be recently ejected from the Midwest compact. Host state responsibilities for the Midwest compact have shifted to Ohio, and Michigan is currently on its own, and is trying to negotiate to locate a disposal site which will take its waste. Whether Ohio will be successful in siting a facility, and whether the Midwest compact will remain intact is at this point an open question.[126]

Another serious concern is the economics of waste disposal. If more radioactive waste becomes deregulated (as the NRC plans), and as volume reduction techniques are applied to reactor waste to reduce costs and meet federal volume limits, and as the number of disposal sites proliferates, the commercial disposal of radioactive waste may become less economically viable. Current plans to build up to 12 new disposal sites may result in underutilized sites and soaring disposal costs. Richard Slember, vice-president of Westinghouse's energy systems business unit, remarked that "[t]he country doesn't need a dozen low-

---

125. Nucleonics 1989a, p. 4.
126. Davis 1991, p. C-1.

level waste sites; it needs three or four good ones."[127] Although the compact legislation was intended to encourage states to join together to prevent a proliferation of sites, the arrangements are politically sensitive and difficult to negotiate.

The average cost for disposal of low-level waste has dramatically increased in the past 20 years. In 1975, for example, unit disposal costs for most Class A low-level waste were about $1 per cubic foot.[128] By comparison, current base disposal charges at the Barnwell site in South Carolina are about $41 per cubic foot,[129] and at the Beatty, Nevada site, unit charges range from $32 to $55 per cubic foot, depending on the radiation levels at the surface of the disposal package.[130]

Expected future cost ranges are huge—from $30 to $40 per cubic foot for a high-capacity (e.g. 230,000 cubic feet per year) conventional shallow-land burial site to $600 to $700 or more per cubic foot for a low-capacity (10,000 cubic feet per year) disposal site utilizing vaults or modular concrete canisters for storage.[131] (By comparison, Barnwell, the largest of the current three sites, accepted about 1.1 million cubic feet of waste in 1989, and Beatty, Nevada, the smallest, accepted 116,000 cubic feet.)[132]

The estimated total development costs for 11 new disposal sites currently under development around the country approaches $700 million. This does not include any estimate for development of proposed new disposal sites for the Rocky Mountain Compact and for the state of Massachusetts.[133]

The bottom line, however, is that, regardless of what the costs turn out to be, the current approach for significant quantities of low-level waste disposal is fundamentally inadequate. Failures may not occur as quickly as they have in the past, but, as the NRC's own regulations admit, that may not make any difference to the future generations who will be left to contend with the long-lived wastes whose hazard will far

---

127. Nucleonics 1989b, p. 9.
128. OTA 1989, p. 144.
129. Chem-Nuclear 1991.
130. US Ecology 1990.
131. OTA 1989, Table 6-6, p. 145.
132. DOE 1990b, p. A-2.
133. Appalachian Compact Users of Radioactive Isotopes (ACURI), excerpt including "Estimated Development Costs for States and Compacts", undated, but apparently later than January 28, 1991.

outlast the regulatory time frame. In that respect, it is similar in spirit to the now defunct EPA standards for high-level waste which limited radiation doses for only 10,000 years for threats which extend to a million or more years.

C. Mill Tailings

Contamination and resultant health-hazard problems exist at uranium milling sites throughout the U.S. Concern about this contamination was sufficient to result in the passage of the Uranium Mill Tailings Remediation and Control Act of 1978, which required cleanup at all current and former U.S. milling sites.[134]

This mill tailings law contains two sections. The first regulates the operation and cleanup of mill sites that are currently under NRC or state licenses, the cost of which is to be paid for by the company which produced the tailings. The second section put the DOE in charge of cleaning up abandoned sites, the cost of which is to be split between the federal government and the affected state government, at a rate of 90 percent/10 percent. These sites are subject to cleanup under DOE's Uranium Mill Tailings Remedial Action Program (UMTRAP).

Under provisions of the law, the EPA was to set general environmental standards for long-term tailings control, and the NRC was to establish licensing requirements for active milling operations.[135]

In 1983, on two separate dates, the EPA issued its standards for cleanup of mill tailings at active and abandoned sites.[136] These standards establish criteria for tailings isolation in lined impoundments which limit radon emission, include groundwater protection requirements, and limit radium concentration in contaminated soils. These standards extend for a time frame of 200 to 1,000 years after the closure of the site.

As of 1985, estimates indicated that the cost of the cleanup at the 28 currently licensed U.S. mills under EPA's minimum national standards would range from $1 billion to $4 billion.[137] Over the last several

---

134. Public Law 95-604, cited in NAS 1986, p. 17.
135. The genesis of these regulations is discussed in SRIC 1985, pp. 110-113.
136. The standards for inactive sites (40 CFR 192, Subpart A) were published in EPA 1983c; active site standards (40 CFR 192, Subpart D) were published in EPA 1983a.
137. SRIC 1985, p. 108.

years, little progress has been made, and at least 13 out of the 25 licensed uranium milling sites which have operated remain unstabilized or only partially stabilized.[138]

As is apparent from the 1,000-year time frame in the regulations, none of the remedial programs contain provisions for protecting future generations for time frames compatible with or longer than the radium-226 half-life (1,600 years) and most certainly not the thorium-230 half-life of 80,000 years. Putting tailings in lined impoundments may prevent leakage in the near term, and perhaps even for the 1,000-year period which the regulations cover. However, it provides little assurance for the great majority of the hazardous life of the principal radionuclides of concern, radium-226 and thorium-230.

We note that the main radioactive hazards from mill tailings, thorium-230 and radium-226, can, in principle, be separated from the tailings and treated as repository-bound long-lived wastes. We here make a preliminary estimate of what this would entail.

The actual costs of separation would vary from site to site and depend on the concentration of these elements in the tailings. Such separation need not be as expensive as might first seem, since there is no need for the "product" (i.e. the separated thorium and radium) to be pure, as is the case for the initial uranium separation or as was the case for the radium industry in the early part of this century. The principal requirement is that substantially all of the radium and thorium be removed from the tailings, and be in a much smaller volume so that they can be disposed of along with other highly radioactive long-lived wastes.

There are on the order of 200,000 curies each of radium-226 and thorium-230 in the roughly 250 million metric tons of uranium mill tailings in this country.[139] The respective weights of these elements are about 200 kilograms and 10,000 kilograms.[140] If the mill tailings are

---

138. DOE 1990d, pp. 130-131. The status of these mills is apparently unchanged since the issuance of a similar report in 1988 (DOE 1988c).
139. Assuming that radium-226 and thorium-230 each exist in the tailings at concentrations of about 800 picocuries per gram. Based on EPA 1983b, Table 3-1, p. 3-6, this is probably an overestimate and therefore conservative. The source lists radium concentrations at licensed mill tailing piles in the U.S. which range from about 200 to 850 picocuries per gram, with many in the range of 400 to 500 picocuries per gram.
140. Based on specific activities of 1 curie per gram for radium, and 0.02 curies per gram for thorium-230.

processed so that the radium and thorium content in the "product" is about one part in a hundred by weight, the total weight to be disposed of as long-lived, repository-bound waste would be about 1,000 metric tons. This is less than one percent of the weight of encapsulated reactor spent fuel, and would therefore be expected to affect only negligibly the size of a repository.

We should note, however, that there are potential drawbacks to this approach, and the idea needs to be considered carefully. For example, the resulting repository-bound radium and thorium concentrates would of course be quite radioactive, with concentrations of about 640 curies per cubic meter, based on the above assumptions.[141] Thus, issues of the risks from waste handling and potential worker exposures would need to be carefully weighed.

Since the total weight of radium and thorium waste in a mixture containing just one part in a hundred of radium-226 and thorium-230 is not huge, it indicates that extraction of these radionuclides is technologically feasible for repository disposal, and therefore should be considered.

---

141. Based on an assumed density similar to that of the mill tailings themselves, i.e. 1.6 metric tons per cubic meter (DOE 1990d, p. 131).

Chapter 4

AN ALTERNATIVE APPROACH

A. Introduction: The Need for an Alternative Approach

The U.S. radioactive waste management program suffers on both ends of the radioactive waste spectrum. The spent fuel and high-level waste problem, although originally slighted, has become the victim of a false sense of urgency which has led to an approach which threatens to risk both environmental and financial failure for the nominal sake of getting a repository on-line and functioning as soon as possible.

On the other hand, the management of "low-level" radioactive waste suffers from a regulatory approach which clearly downplays the potential hazards for the more dangerous categories of low-level waste.

The common thread of these two flawed approaches is a perspective which has falsely dichotomized the two types of waste. The near exclusive focus had been on the spent fuel and reprocessing wastes, while other radioactive wastes—some of which are comparably hazardous—have been relegated to the background with the innocuous sounding, but sometimes misleading term, "low-level."

At the same time, wastes which are designated as "transuranic" and are similar in some respects to the more dangerous and long-lived categories of "low level" waste, are to be consigned to a repository. However, there is no stated policy for the majority of transuranic wastes which remain buried just below the surface, since the one deep repository touted as the transuranic waste "solution" can only accommodate a fraction of existing and projected transuranic wastes. Meanwhile, there is no policy for protecting the environment from the long-lived components of uranium mill tailings beyond 1,000 years, despite the fact that these hazards will persist for hundreds of thousands of years.

This Alice-in-Wonderland situation clearly calls out for a more sensible approach to radioactive waste management. Such an approach would rely on a hazard-based classification system rather than one based on an arbitrary boundary drawn at the surface of irradiated fuel rods. In constructing an alternate approach, it should be kept in mind that the long-term disposal problem is defined primarily by the longevity of the waste hazard. That is, the principal technical difficulties associated with disposal relate to the ability or lack thereof to assure long-term containment. Consequently, it would make sense to determine the disposal method (as opposed to interim management or storage) of a given class of waste according to the half-lives of its longest-lived radionuclide constituents which are present in appreciable quantities. It is also necessary to institute policies to minimize the production of such long-lived wastes.

A sensible integrated approach to waste management and classification will not be of any use, however, if the program for managing and permanently disposing of this waste is fundamentally flawed and inadequate. Our alternative approach therefore must also include proposals for re-structuring the existing program.

There are three principal elements of our comprehensive alternative approach:

1) A reclassification of wastes according to longevity of waste. All wastes containing significant amounts of long-lived radionuclides would be slated for disposal via the most stringent long-term management option. According to the current conception, this would be emplacement in a deep geological repository.

2) A deliberate restructuring of current repository programs to put them on a scientifically and environmentally sound basis. This includes the development of contingency plans for options other than the land-based repository approach (such as sub-seabed disposal and, possibly, transmutation for some existing reprocessing wastes) in order to ensure that the best approach for long-term isolation is adopted. Such restructuring would entail the elimination of the current unrealistic and overly rigid timetable for characterizing Yucca Mountain and building a repository there by 2010,

and an end to DOE efforts to open the fundamentally flawed Waste Isolation Pilot Plant (WIPP) repository for transuranic wastes.

3) A restructuring of interim management plans to accommodate the likely need for extended onsite storage for many decades beyond what the DOE now foresees. On the commercial side, this includes:

- planning to provide capacity for extended storage of spent fuel and some portions of the low-level waste, for up to 100 years at the reactor site which generated the waste.

- a deferral of the decommissioning of shut-down nuclear power reactors for up to 100 years until a long-term waste management system is operational for the spent fuel and long-lived decommissioning wastes.

On the military side, this restructuring includes onsite retrievable storage of transuranic and long-lived low-level wastes in carefully monitored facilities that comply with all appropriate environmental regulations (including all those, such as RCRA, which would apply to a commercial firm engaged in similar waste-management activities).

The details of this alternative approach are laid out in the sections below. In these sections we assume, for the sake of discussion, that a land-based repository will be the method of disposal for long-lived radioactive wastes. The approach we recommend does allow, however, for other approaches to be developed which might be an improvement on or a complement to the land-based repository one.

B. Component 1: Waste Reclassification

The disposal method adopted for a given type of waste should depend principally on the longevity of the hazard presented by the waste. In general, the longer the hazard, the more secure must be the disposal method, and the more stringent must be the environmental standards for long-term isolation. For example, wastes bound for a deep geological repository should not be defined by the process which produced them (which does not necessarily bear a direct connection to their health or environmental risk), but should instead be defined as including essentially all long-lived wastes.

Such a definition should be at least inclusive enough to include both strontium-90 and cesium-137 as long-lived wastes, provided appropriate concentration criteria are satisfied. Considerable additional careful study and democratic debate will be required before these criteria can be determined.

An analogue to this sort of approach already exists in the U.S. government's definition of transuranic wastes. As previously discussed, these wastes are defined by the DOE as including all waste material that contains transuranic elements with half-lives greater than 20 years and whose concentrations are 100 nanocuries per gram or greater.[1] Although the interim management and handling procedures vary considerably for transuranic wastes depending on the level of penetrating radiation emitted,[2] the principal determinant of its permanent disposal methods is the long-lived transuranic content. Although most of these wastes have relatively low levels of radiation in terms of intensity and penetrating power, the very long-term nature of the hazard typically posed by transuranics (especially such elements as plutonium-239), and the high damage potential from inhalation and ingestion, is what requires them to be disposed of in a deep geologic repository according to standards similar to those set for high-level waste and spent fuel disposal.

According to our proposed approach, dilution of wastes to circumvent concentration criteria or to avoid putting them in a repository would be prohibited.[3] On the contrary, one operating principle should be to concentrate long-lived radionuclides wherever possible, so that as much of the long-lived material as possible can be sent for deep geologic disposal.

---

1. DOE 1990d, p. 75.
2. Transuranic-waste containing beta, gamma, or neutron emitters sufficient to result in doses greater than 200 millirems per hour is designated as "remote-handled" waste because of the precautions which must be taken to protect workers who handle it. About 2.4 percent of current retrievably stored transuranic waste is remote-handled waste, while the remainder is classified as "contact handled" transuranic waste. (Most transuranics are alpha-emitters, many with half-lives in the hundreds or thousands of years. Alpha radiation is not very penetrating, but its biological hazard is relatively high if it is emitted internally following inhalation or ingestion of alpha-emitting radionuclides.)
3. This sort of prohibition already has precedent in federal hazardous waste regulations, which prohibit hazardous waste from being removed from regulatory purview simply through dilution with a non-hazardous waste. [Resource Conservation and Recovery Act (RCRA) regulations, at 40 CFR Part 261, as described in EPA 1986, pp. III-14, III-15.]

There are several important consequences of such an approach for radioactive waste management policy:

- significant quantities of commercial radioactive wastes now classified as low-level waste and permitted to be disposed through shallow-land burial would be re-classified as repository-bound wastes.

- similarly, some quantities of long-lived radioactive waste from DOE nuclear weapons activities now planned for disposal in shallow land burial as low-level waste would also be re-classified as repository-bound waste.

- The expected repository capacity requirements would consequently increase to allow for the increased volume going to a repository.

We make some preliminary estimates of the magnitude of some of these changes based on an analogy to the approach taken in Sweden. Specifically, we estimate the volume of low-level waste from commercial reactor operation that would go to the repository based on this alternative approach. We can then estimate the consequences this will have on the repository capacity requirements.

We do not at this time have adequate information to review the volume implications for reclassification of DOE low-level wastes. Based on available information, however, including all long-lived DOE wastes may increase required capacity volume by as much as or more than the volume additions from commercial reactor low-level waste.[4] There are also considerable quantities of transuranic wastes not slated to go to the WIPP repository if, indeed, any wastes go there at all. The inclusion of these wastes and future wastes like them from the nuclear weapons complex could also considerably increase repository space

---

4. For example, if the increase in repository volume requirements were proportional to existing annual low-level waste generation rates, the contribution to repository volume requirements from re-classification of DOE wastes would be several times the contribution from commercial reactor low-level waste. This is simply a consequence of the fact that DOE waste generation rates in 1988, for example, were four to five times that of commercial reactor generation rates. However, since the character of much of DOE low-level waste is not necessarily the same as that from commercial reactor low-level waste, the situation is more complicated.

requirements.

Before going into the details of the U.S. situation, we introduce as an example the approach taken in Sweden towards waste classification.

The Case of Sweden

By comparison to the U.S. classification system for radioactive wastes, the conceptual approach adopted by Sweden is a more rational one.[5] For example,

> [Sweden] classif[ies] ... radioactive waste in two different ways, either as low-level, intermediate-level and high-level with regard to the needs for shielding and cooling during handling and storage, or as short-lived and long-lived with regards to the demands on the final disposal method.[6]

Accordingly, the Swedish repository program for long-lived wastes anticipates disposal of numerous low-level and intermediate-level wastes containing long-lived radionuclides along with its spent fuel. Table 11 shows the breakdown of Swedish waste according to source and destination.

As the table indicates, the waste disposal system has two main components: an "SFL" repository for long-lived wastes, and an "SFR" repository for shorter-lived wastes. Examining the expected waste volumes for the SFL repository, we see that only 26 percent of the final volume is due to spent fuel. The remainder consists of various wastes which would be considered "low-level" in the U.S.: operating waste resulting from storage and encapsulation activities for spent fuel, decommissioning wastes from the facilities used for this purpose, and other waste (mostly core components and reactor internals) from reactors. Most interesting to note is that fully 40 percent of the volume (19,700 cubic meters) in this long-lived waste repository will be non-spent fuel wastes directly from reactors. This 19,700 cubic meters represents over 11 percent of the non-spent fuel waste volumes from power reactors.

---

5. We note, however, that we have not examined the details of the Swedish implementation of the program, and make no judgement here in this regard.
6. Hans Forsstrom, Director, Swedish Nuclear Fuel and Waste Management Company (SKB), Personal Communication to Scott Saleska (23 May 1991).

Table 11
RADIOACTIVE WASTE IN SWEDEN THROUGH 2010

Area	Contents	Volume (cubic meters)	Percent
	Waste to Swedish "SFL"* Facility		
SFL-1	Originally for virtified reprocessing waste no longer applicable		
SFL-2	Encapsulated Spent Fuel	12,800	26%
SFL-3	Operating Waste From CLAB (an MRS-like facility)	4,100	8%
	Operating Waste From Encapsulation	900	2%
	Waste from Studsvik (an experimental facility)	1,500	3%
SFL-4	Operating Waste from CLAB	700	1%
	Decomm. Waste from CLAB and Encapsulating Facility	8,900	18%
	Transportation Containers	600	1%
SFL-5	Core Components & Reactor Internals	19,700	40%
TOTAL SFL WASTE			100%
	Waste to Swedish "SFR" Facility		
SFR-1	Operating Waste from CLAB	2,500	1%
	Waste from Studsvik Experimental Facility	14,000	7%
	Operating Waste from Nuclear Plants	72,800	38%
SFR-2	Originally for Core Components	replaced by SFL-5	
SFR-3	Decommissioning Waste from Reactors	100,000	51%
	Decommissioning Waste from Studsvik	4,000	2%
TOTAL SFR WASTE		193,300	100%
TOTAL SFL & SFR WASTE		245,500	

Source: SKB 1990, Appendix 1.
* Final repository for long-lived waste–an assumed 500-meter-deep geologic repository.
** Final repository for operating waste–located 60 meters under Baltic Sea.

The other 89 percent (172,800 cubic meters) of operating and decommissioning wastes from nuclear reactors are expected to go to the SFR repository for short-lived wastes (72,800 cubic meters into SFR-1, and 100,000 cubic meters into SFR-3).

Note that even these latter wastes, which are considered relatively "short-lived," are being disposed of in a facility which is being mined from rock 60 meters below the bottom of the Baltic Sea. This is not as deep as a typical long-lived waste repository such as the SFL Swedish facility or what is planned for the U.S., but is much more extensive than the conventional land burial practiced in the U.S. for low-level waste. Sweden is not the only country disposing of low-level wastes in a deep underground repository, or something similar to it. Finland plans to dispose of similar wastes from its nuclear power plants in repositories about 300 feet beneath each plant (although not quite qualifying for "deep," this is much deeper than envisioned for the U.S.). England proposes to dispose of low- and intermediate-level wastes in a repository 1,000 feet underground. And West Germany disposed of some low-level waste in the Asse Salt Mine between 1967 and 1978.[7]

An additional notable feature of the Swedish program is that Sweden plans to phase out nuclear power by the year 2010.

Reclassification of U.S. Commercial Low-Level Wastes

Ideally, in order to apply a reclassification approach, volume estimates would be based on specific technical concentration criteria for each radionuclide, analogous to the 100 nanocuries per gram concentration limit for transuranics. These concentrations would then be compared to the characteristics of existing wastes to determine what portion of these wastes should be reclassified.

It is beyond the scope of this study to suggest specific numerical limits on a per-radionuclide basis for wastes in the U.S. due to the great variety of wastes in the nuclear weapons complex which must also be considered in an integrated program, in addition to commercial wastes. We here make a preliminary, order-of-magnitude estimate based on the Swedish waste categorization system in order to evaluate the potential effect on repository size requirements of reclassification of wastes originating from the commercial nuclear power industry.

As was shown in Table 11 in the previous section, the waste volume to be disposed in the Swedish long-lived waste repository (SFL) includes core components, reactor internals, and low-level waste from encapsulation of the spent fuel. These are wastes directly associated with the commercial nuclear fuel cycle, most of which would be disposed of as low-level wastes under the current system used in the U.S.

Assuming that a restructured waste classification system in the U.S. would assign roughly the same proportion of low-level wastes for long-lived repository disposal as is the case in Sweden, the repository-destined low-level waste volume can be calculated based on the ratio

---

7.   OTA 1989, p. 127.

of low-level waste to spent fuel volumes in the Swedish SFL repository.[8]

The result of this calculation is that, assuming the DOE's no-new-orders scenario prediction of a total of about 86,800 metric tons of spent fuel generation, about 225,700 cubic meters of low-level waste associated with commercial power reactor operation will have to be emplaced in a deep underground repository along with this spent fuel.[9]

This result, along with the assumptions about packaged waste volume and area requirements for spent fuel and high-level waste, are depicted in Table 12 in the section below on "New Repository Capacity Requirements for Long-Lived Waste." This table also shows the anticipated repository area requirements for additional waste volumes, which is also discussed in that section.

New Repository Capacity Requirements for Long-Lived Waste

Estimating the additional repository space required due to the additional volume from reclassified low-level waste depends on the spacing of the wastes in the repository. Spacing of spent fuel and high-level wastes is limited by the heat load limitations imposed by the repository design and environmental standards. This limitation is referred to as the "areal power density." For spent fuel, the currently assumed areal power density is 57 kilowatts per acre. For reprocessing high-level waste, it is 83 killowatts per acre.[10] This limitation, combined with the power levels generated by a typical canister of waste emplaced, determines the area requirements of the repository. The

---

8. This ratio is based on the numbers in Table 11. A factor of 3.6 in the packaged spent fuel volume of 12,800 is due to the Swedish spent fuel containers, so the actual volume of spent fuel is 3,556 cubic meters. Thus, the ratio is [19,700 (core components) + 900 (encapsulation)] / [3,556 (spent fuel)], or 5.79. This implies that for each cubic meter of unpackaged spent fuel in the repository, there will be 5.79 cubic meters of packaged reactor-associated low-level wastes in the repository.

9. This calculation is as follows: 86,800 metric tons of approximately 1/3 BWR spent fuel, and 2/3 PWR spent fuel has, according to DOE data (DOE 1990d), a volume of about 38,960 cubic meters. Multiplied by the low-level waste factor of 5.79, this gives about 225,700 cubic meters of low-level waste.

10. Mansure 1985, pp. 6, 9.

specific limits are currently the subject of considerable debate, but the area requirements due to the above assumptions are incorporated into Table 12.

Table 12 RADIOACTIVE WASTE DISPOSAL SPACE REQUIREMENTS					
Waste Type	Tons	Volume	Packaged Volume	Density (vol/ acre)	Repository Area Req (acres)
Spent Fuel Metric Tons Uranium					
Boiling WAter Reactor	28,919	13,973[a]	NA[b]	NA[c]	NA
Pressurized Water Reactor	57,838	24,985[a]	NA[b]	NA[c]	NA
TOTAL	86,757	38,958	83,120	36.1[c]	2,300
Reprocessing Waste					
Military Wastes	8,875[d]		19,432[e]	193.3[f]	101
West Valley Wastes	640		328	193.3	2
TOTAL	9,515		19,761	193.3	103
Reclasssified Commercial LLW (incl. core comp., reactor internals, encapsulated wastes)					
LLW high-area estimate			225,712[g]	193[h]	1,167
LLW low-area estimate			same	1,600[i]	141
TOTAL REPOSITORY AREA REQUIREMENTS				LOW: HIGH:	2,545 3,570

Sources and Notes:
(a) Spent fuel volume factors are 0.483m^3/MTU (BWR), and 0.432m^3/MTU (PWR) (DOE 1990d). Note that federal law (the 1982 Nuclear Waste Policy Act) currently imposes a 70,000 metric ton limit on the total amount of waste the first repository can receive.
(b) Assumes intact disposal of most spent fuel in hybrid canisters containing 3 PWR and 4 PWR assemblies each, (DOE 1989b, p. B-11).
(c) Spent fuel volume density derived from assumed areal power density of 57 kW/acre 3kW per canister (Mansure 1985, p. 6),and canister volume of 1.9 m^3 (DOE 1989b, p. B-12).
(d) Metric tons of spent fuel reprocessed is from DOE 1990a, p. 8.
(e) Based on an assumed 17,750 canisters from military sources, 300 canisters from West Valley,(DOE 1990a, p. 8), and canister volume of 1.09 m^3 (DOE 1989b, p. B-13).
(f) Reprocessing waste volume density derived from an assumed areal power density of 83 kW per acre, canister volume of 1.09 m^3 and canister 0.47 kW/canister. (Mansure 1985, p. 9)
(g) Packaged low-level waste volume assumed Swedish (Packaged low-level waste)/(Unpackaged Spent fuel) volume ratio of 5.79, and rough equivalence between unpackaged U.S. and Swedish spent fuel assemblies, and between Swedish and U.S. low-level waste packaging volume. Does not include additional low-level waste volume from DOE/military sources.
(h) High-area estimate derived assuming same emplacement density as for high-level reprocessing wastes.
(i) Low-area estimate derived assuming same emplacement density as for transuranic waste emplacement in WIPP.

For repository-destined low-level waste, we expect a wide range of thermal power densities, ranging perhaps from the thermal density on the order of that of vitrified high-level reprocessing waste down to relatively insignificant levels. In order to estimate additional repository volume requirements, we assume a range of plausible emplacement densities. This range is bounded on the low end by the emplacement density of vitrified reprocessing waste (193 cubic meters per acre), and on the upper end by the emplacement density anticipated for transuranic wastes in WIPP (1,600 cubic meters per acre).[11] This range results in area requirements ranging from about 140 to 1,200 acres of repository space.

The repository requirements for a "no new orders" scenario could then be in the range of about 2,500 to 3,600 acres, excluding the long-lived components of military low-level wastes and transuranic wastes for which there is currently no designated repository. When these two categories of waste are included, the upper limit of repository space may be significantly larger.

C. Component 2: Restructuring Long-Term Waste Management [12]

The present process of site selection and characterization has been thoroughly compromised both technically and institutionally. It is essential to abandon it. This includes both the program to characterize Yucca Mountain as the site for high-level waste disposal and the present plan to use the WIPP site in New Mexico to dispose of a portion of transuranic wastes. The false sense of urgency which has pervaded the program needs to be abandoned by the utilities, by the federal government and by those who have felt that a law mandating a repository would somehow reduce the risks of nuclear weapons proliferation by eliminating the option of reprocessing. The current repository program does not eliminate that risk.

Further, the present NRC regulations and corresponding state regulations for "low-level waste" should be scrapped. State or compact-level regulation and management should apply only to short-lived wastes. As we have proposed, long-lived wastes from all sources

---

11. DOE 1990f.
12. Much of this section is adapted from Makhijani 1989, pp. 93-101.

should be integrated into the repository program or its equivalent alternative.

Since there are both technical and institutional flaws in the current program, a restructured program must address both aspects.

Technical Aspects

The principal technically oriented components of developing a long-lived waste management solution should include: time, flexibility, redundancy, contingency planning, and waste minimization. These five components are discussed below.

The first component is time. Our present knowledge of geology and climate is insufficient to enable us to predict the performance of repositories over time scales on the order of a million years. Any assessment depends on a variety of assumptions, such as stability of climate and extension of past geological phenomena into the long-term future, which cannot at present be predicted with sufficient confidence to ensure a reasonable degree of environmental protection for the hazardous life of the waste. This means that considerable uncertainties remain even after the best efforts, not even accounting for the additional uncertainties due to the problems with DOE's flawed analyses.

A number of factors, many of them related to environmental protection, are propelling science and society in the direction of making substantial efforts to understand climatic and geologic change. For instance, a crucial part of the motivation of NASA's "Mission to Planet Earth" is understanding of the interaction between atmospheric composition, temperature, and the patterns of climate and climate change on Earth. Our understanding of climatic and geologic processes is likely to increase dramatically in the next decades due to such efforts. This will help considerably in addressing the uncertainties associated with site selection, characterization, and performance assessment.

The main point here is that the present knowledge is inadequate, and that time needs to be made available to do the technical job right, rather than under the constraint of arbitrarily imposed deadlines.

The second component is flexibility. We mention this here with particular reference to the recent report by the National Academy of Sciences Board of Radioactive Waste Management which criticized the rigidity of the present program. The NAS Board recommended that a more flexible approach be taken which could anticipate and respond to surprises that arise in the development of a repository. Such an approach should not, according to the Board, be bogged down in rigid and uncompromising development plans that had been specified in minute detail before the technical site investigations had seriously begun. The NAS Board summarized its recommended flexible approach in three principles:[13]

- Start with the simplest description of what is known, so that the largest and most significant uncertainties can be identified early in the program and given priority attention.

- Meet problems as they emerge, instead of trying to anticipate in advance all of the complexities of a natural geological environment.

- Define the goal broadly in ultimate performance terms [e.g,. in terms of individual doses to maximally exposed individual], rather than immediate requirements, so that increased knowledge can be incorporated into the design of a specific site.

Such an approach makes sound technical sense, and should be used, especially when knowledge has developed to the point where development at a specific site or sites is underway. However, such calls for flexibility should not be interpreted as a license to relax environmental protection standards. Some parts of the Academy Board's report seem to imply just such license. Our contingent endorsement of the NAS recommendation regarding technical flexibility has institutional management ramifications which we address in the next section on "Institutional Aspects."

The third component of a well-designed long-term disposal solution is technical redundancy to ensure isolation of wastes. The current U.S. program is lacking in this area. For example, it relies too much on geologic isolation, failing to take adequate advantage of the potential of engineered barrier systems. The Swedish nuclear waste program, by contrast, has studied the engineered barrier potential in detail.

A spent fuel canister with a much longer design life would help reduce the need to rely primarily on geologic isolation. The Swedish approach to long-term waste management puts considerably more

---

13. NAS 1990, p. 7.

reliance on the performance of the waste canister as an isolation mechanism than the U.S. approach. As an ad hoc National Academy of Sciences Panel noted in 1984:

> The Swedish plan differs from most others in its heavy reliance on engineered barriers, specifically thick-walled copper canisters to enclose the spent fuel rods, surrounded by buffers of compacted bentonite.[14]

The Swedish estimate for isolation which would be achieved by the specially designed canisters, made by the Swedish Corrosion Institute, was on the order of 1 million years. Although it is difficult to predict over such long time frames, an approach which aims for such a goal for an engineered barrier, in conjunction with geologic isolation, is a sound one by virtue of its technical redundancy.

In brief, our recommendation is that the NRC regulation based on an assumed canister life of only 1,000 years and the complete disintegration of the canister to 100,000 years should be scrapped. The standards for a redundant repository system should impose requirements on both the geologic and engineered barrier components as if they each were the main guarantor of isolation. In other words, there should be equal weight on each component, and regulations should be equally stringent for each.

The fourth component of a technically robust long-term waste disposal program is the use of contingency planning. Although geologic disposal in a deep repository has been the basis for planning for many years, the U.S. waste program should plan for the contingency that, after additional research, it may not turn out to be the best option. Research should therefore continue on other options, notably sub-seabed disposal. Sub-seabed disposal involves emplacement of nuclear waste cannisters in geologic formations under the ocean floor in areas thought to be geologically stable.[15] Numerous issues such as transportation risks on land and sea, and questions of international jurisdiction, regulations, and standards would have to be faced. In recommending that research continue on this option, we are not suggesting that it is an easy "solution" to this problem, but rather that it may be a potential alternative in a very difficult situation.

---

14. NAS 1984, p. 2.
15. Hollister 1981; Berlin 1989, p. 294-296.

The final component of such a program should be waste minimization for long-lived radionuclides. This has significant implications for both nuclear power and nuclear weapons production. In view of the fact that there is currently no viable scheme for disposal of these wastes, and given their immense implications for the health of future generations, waste reduction to the maximum extent feasible is essential. A revamping of policies regarding classification and disposal should therefore also consider various incentives and disincentives for generators of waste to reduce or eliminate the production of long-lived radionuclides. Among these incentives should be full cost internalization and removal of limitations on liability including those provided by legislation such as the Price Anderson Act.[16] Evidently, the expanded use of alternative energy sources and the implementation of vigorous energy conservation programs would help reduce reliance on nuclear power and also reduce the magnitude of the problems associated with long-lived radioactive wastes. Such measures would also be helpful in dealing with the need to reduce the emissions of carbon dioxide, an important greenhouse gas central to the global climate change problem. A thorough examination of these issues, however, obviously involves the broader realm of energy policy and is beyond the scope of this book.

Institutional Aspects

The Department of Energy has repeatedly failed to conduct site selection and related activities with the scientific, technical and institutional integrity needed for a program with such serious implications for the health of future generations. The DOE's principal emphasis has been on promoting nuclear power and in helping nuclear utilities to get rid of the spent fuel from power plant sites, even at the cost of such integrity. Likewise, its principal emphasis in the weapons program has been on weapons production, even at the cost of serious damage to the sites and of similarly compromised science.

Indeed, one of the principal reasons for the existence of the overly rigid, inflexible approach that now exists which, as discussed above, was the object of strong criticism by the NAS Board on Radioactive

---

16. The Price-Anderson Act is a law which imposes a limit on the liability which can be incurred by a company in the event of a nuclear power accident, thus artificially reducing overall nuclear power operating costs by substantially reducing insurance premiums.

Waste Management, is the DOE's consistent record of failure in dealing with the radioactive waste problem. This has led the public, Congress, other governmental agencies, and specially appointed independent panels to take a heavy-handed, highly prescriptive approach, tightly overseeing DOE activities in this area in order to help ensure that past failures do not recur. Of course, such an approach only introduces a new form of rigidity which the NAS Board rightly criticized as a source of more potential for failure.

One of the things we agree with the Board on is that the waste program needs more flexibility. However, the DOE's penchant for evading regulations means that the intent of flexibility may be subverted and defeated—instead of a sounder program, we may get an even more compromised one. Thus, we are faced with the paradoxical situation in which the oversight necessary to impose minimal accountability on the DOE has produced an intractable rigidity, but a program with the needed degree of flexibility probably cannot be implemented with integrity by the DOE.

The only apparent option to restore confidence, credibility, flexibility, and competence is to replace the DOE with an independent waste management authority.

There are several aspects to the institutional questions associated with high-level wastes that must be factored into a decision about how these wastes are to be managed. The first is the separation of the regulation of the interim, onsite management from the long-term research aspects. The second is the creation of a process and an institution that will be free of conflicts-of-interest and that will have the scientific integrity, technical capability, independence, managerial competence, and public accountability to carry out what is a very difficult task.

We discuss aspects of interim management through onsite storage of wastes in the next section. The institutional arrangements for addressing the long-term management problem need to be considerably different from those needed for interim management. The long-term problem is, at present, mainly a research and development question. One option that is *unlikely* to work is simply breaking off the Office of Civilian Radioactive Waste Management (the DOE office in charge of the current program) from the DOE and making it a separate, supposedly independent body. The commitment of the present office to nuclear power and the large momentum it has towards characterizing and building a specific repository would make it inappropriate to address what would at present be essentially a scientific task.

A body such as the National Science Foundation, or an ad-hoc agency with a board of directors which includes appropriate federal and state agencies would be more appropriate to the task. The agency would have to operate with a clear agenda for research and development which would *exclude* final site selection, but would include study of all possible media for disposal which would minimize risk to future generations. These would include land-based geologic disposal as well as sub-seabed disposal.

Whatever the specific make-up of the new agency, its structure and operation will have to ensure the highest standards of scientific integrity, public accountability, and financial responsibility. Its goals must be in the direction of minimizing risk to future generations from nuclear wastes and it must, therefore, be free of the kind of conflict-of-interest favoring nuclear power or nuclear weapons production which has been a problem with the present DOE-managed program.

D. Component 3: Interim Management Via Extended Onsite Storage

In order to accommodate the needs of a restructured development program for long-term waste management, extended onsite storage will be needed for various categories of waste. Associated with this will be the need to defer the decommissioning of nuclear power reactors as long as the stored long-lived radioactive waste remains on site. Like the issue of long-term management, interim management of long-lived radioactive wastes has both technical and institutional aspects.

Technical Aspects

Onsite storage of long-lived radioactive wastes will have to be expanded and extended at both commercial nuclear power sites and at DOE weapons sites.

Regarding commercial nuclear wastes, capacity should be provided for extended storage of spent fuel, as well as the long-lived portions of low-level waste, for up to 100 years at the reactor site which generated the waste. Spent fuel can be stored for extended periods using the dry cask storage technologies which we have discussed in Chapter 3.

One hundred year onsite storage would also necessarily entail a deferral of the decommissioning of commercial reactors when they are shut down. Utilities should plan to defer decommissioning for up to 100 years, or until a long-term waste management system is operational for the spent fuel and long-lived decommissioning wastes that are stored at the reactor site.

DOE nuclear weapons sites will also require onsite retrievable storage of high-level, transuranic, and long-lived low-level wastes in carefully monitored facilities that comply with all appropriate environmental regulations (including all those, such as RCRA, which would apply to a commercial firm engaged in similar waste-management activities).

A program of extended onsite storage, in addition to being integral to a restructuring of long-term waste management, is likely to have a number of other incidental advantages as well, including:

- A reduction in the temperature of highly radioactive wastes due to radioactive decay. This will enhance safety and enable a wider variety of canister designs and geologic media to be considered. It will also enable a larger quantity of waste to be disposed of in a repository of a given volume;

- A concomitant reduction in transportation hazards associated with highly radioactive wastes due to the decay of the shorter-lived radioactive components such as krypton-85, strontium-90, cesium-137, and plutonium-241; and

- A reduction in the costs and dangers of decommissioning nuclear reactors, provided adequate monitoring and security is maintained.

We should emphasize that onsite storage is not a long-term solution, and cannot take the place of a long-term method of managing nuclear waste, such as that envisioned for an appropriately structured repository program.

There are a number of difficulties with onsite storage that make it less than ideal:

- The need for continuous maintenance, monitoring and surveillance, which cannot be guaranteed for the hazardous life of the waste;

- The need for security to prevent access to wastes by intention or accident; and

- The lack of assurance that the storage facilities used today will themselves be safe for the very long term. In addition, interim storage facilities are not without potential safety or environmental problems. For example, spent fuel casks may be subject to the release of gaseous radionuclides due to rough handling,[17] and at some nuclear sites there may also be concerns arising from seismicity hazards.

Despite these concerns, however, onsite storage is likely to be the least dangerous interim option until a long-term disposal method is developed.

Institutional Aspects

As far as the interim extended onsite storage of spent fuel is concerned, the Nuclear Regulatory Commission should continue to regulate and license any needed additional storage facilities. Such regulation should also apply to other long-lived wastes generated at nuclear power plant sites, whether these arise from routine operations or decommissioning. Since, as far as commercial spent fuel is concerned, this will shift some costs from the MRS-repository system to reactor sites, we propose that funding for this additional spent fuel storage come from the Nuclear Waste Fund.

As far as the interim arrangements for radioactive wastes (including "low-level", "transuranic", and "mixed" wastes) at the nuclear weapons plants are concerned, the EPA, NRC, and their state-level counterparts should have the authority to regulate the management and interim storage of these wastes, including "low-level" wastes, "mixed" radioactive and hazardous wastes, and "transuranic" wastes.

The question of the much smaller amount of "low-level" wastes generated at locations other than nuclear power plants and weapons plants is more complex. There is an immense diversity of sources which produce such wastes ranging from hospitals to oil-field operations. The reclassification of "low-level" wastes recommended here will necessitate the reconsideration of the institutional issues connected with the management of these wastes. Extended onsite storage of long-lived "low-level" wastes at these sites—which number in the

---

17. DOE 1989d, p. I-95.

thousands—may not be possible or desirable. Interim strategy arrangements for such wastes, along with policies for waste minimization, belongs in the province of joint federal-state regulation. Nuclear power plant sites may be appropriate places for such interim storage arrangements since the vast majority of "low-level" wastes in terms of both volume and radioactivity are generated at these sites. Short-lived, low-level wastes would continue to be regulated by states and the NRC.

E. Cost-Benefit Considerations: Current Approach Versus the Alternative Approach

The implementation of the proposed alternative approach to radioactive waste management will have a number of cost implications for the program. Although a detailed integrated comprehensive cost analysis is beyond the scope of this study, it is possible to indicate what some of the changes will be by considering the various components of waste management. Based on considerations raised by existing studies and data, we can draw qualitative conclusions about the direction of cost changes for each of these components. The components we consider here are:

- Integrated costs associated with commercial spent fuel management and disposal in a repository system;

- Long-lived commercial "low-level" waste disposal costs;

- Commercial reactor decommissioning costs; and,

- Military high-level and "low-level" wastes.

Spent Fuel Storage and Disposal Costs

Taking the time to do the long-term disposal job right means that the costs of onsite storage which would have to be undertaken in the interim would of course be greater than now envisioned. However, there is good reason to believe that total-system costs will not vary significantly even with moderately large variations in the opening date for a repository. For example, a detailed cost analysis by the MRS

Review Commission examined the impacts on life-cycle costs of varying repository availability over 20 years, from 2003 to 2023.[18] The relevant results of this are given in Table 13, where costs are given in both constant dollars and discounted to present value at a four percent discount rate.[19]

The actual 2003 date is now moot, since the DOE's earliest target date is 2010. However, the table is helpful in indicating the relative effects of moving the repository opening date. This table shows that, as expected, at-reactor storage costs rise dramatically as the repository opening date is slid back 20 years, from $2.3 billion to $6.6 billion, a 186 percent increase. However, total-system costs rise only about 13 percent, from $24.8 billion to $28 billion, due to declines in other categories.

Even this slight increase, however, is likely to overstate the financial effects of moving the date of repository opening, for at least two reasons. For one, it is stated in nominal dollars. If costs are discounted to present value, this increase is likely to be reversed in favor of a cost decline. Table 13 indicates the effects of the Commission's use of a 4 percent annual discount rate: a net cost decline from $11.6 billion to $7.4 billion in discounted dollars as a result of shifting the repository opening date from 2003 to 2023. For the sake of conservativism, we would hesitate to endorse the MRS Commission's use of a discount rate of as much as 4 percent, even though it is relatively moderate, because it extends over such a long time period. Even at a small discount rate like 1 or 2 percent, however, costs are likely to remain roughly the same or decline slightly.[20]

---

18. MRSRC 1989.
19. Making a choice today between several alternative scenarios which entail future spending usually involves comparing each scenario in terms of its "present value" that is, the amount of money that would have to be set aside today to cover future expenditures as they are incurred. Determining the present value involves taking account of the fact that money set aside today will earn compound interest, thereby reducing the amount necessary to cover future costs. The farther in the future a given cost is, the lower the present value of that future cost. Present value discounting of future costs is a standard methodology used for choosing between alternative future cost streams in a consistent manner, and is used even when the money is not actually going to be set aside today. As the MRS Commission study notes, "opinions differ about the appropriate rate of discount to be used in comparing alternatives through time, but most experts agree that discounting should be done." (MRSRC 1989, p. 71.) The DOE's studies, however, are traditionally reported in undiscounted, constant dollars.
20. Because the individual cost components were not available to us on an annual basis, we were not able to perform this exact calculation.

Table 13 COMPARISONS OF TOTAL-SYSTEM LIFE-CYCLE COSTS (Costs in billlions of constant 1989 dollars)[*]			
Cost Component	**Repository Availability Date**		
	2003	2013	2023
	---- nominal value $ (present value $[**]) ----		
[Storage at operating reactors	1.1 (0.6)	2.4 (1.2)	3.0 (1.3)]
[Storage at shut-down reactors	1.2 (0.4)	2.7 (0.6)	3.6 (0.7)]
TOTAL at reactor storage	2.3 (1.0)	5.1 (1.8)	6.6 (2.0)
Development and Evaluation	9.0 (6.3)	9.0 (4.8)	9.0 (3.8)
Transportation	3.7 (1.0)	3.3 (0.6)	2.9 (0.4)
Repository	9.7 (3.2)	9.2 (2.0)	9.4 (1.3)
TOTAL	24.8 (11.6)	26.6 (9.2)	28.0 (7.4).

Source: MRSRC 1989, p. 73.
* Assumes no MRS.
** Costs in parenthesis are discounted to present value at an annual discount rate of 4 percent.

The second factor is due to the passage of events since the release of the MRS Commission's report—namely, the announcement by the DOE that the earliest possible repository opening date is now 2010. As we mentioned above, all of the cost increase is due to increased cost of onsite storage as the repository opening date is slid into the future. However, the largest portion of this increase occurs due to a repository opening delay from 2003 to 2013 (which causes a 122 percent increase in storage costs, from $2.3 billion to $5.1 billion), and not from 2013 to 2023 (which results in a 29 percent increase in storage costs, from $5.1 billion to $6.6 billion). This seems to indicate that most of the increase in storage costs is due to capital outlays for actual construction of expanded storage facilities, rather than ongoing maintenance costs. This means that the largest portion of the cost increase is already inevitable, due to the fact that even according to the DOE, a repository will not be available before 2010.

The MRS Commission analysis is not strictly applicable to our approach, of course, and cannot completely substitute for it. In particular, it does not consider longer slippages of the repository timetable. However, it is indicative of the effect of sliding the repository opening date back in time. It is also useful in identifying the rather large trade-off in costs that occurs between onsite storage and other components of the program. This supports our recommendation that the

Nuclear Waste Fund be allowed to fund additional onsite storage costs, since this becomes an integral part of the waste management and disposal program, and represents a cost trade-off for reductions in other areas.

To the extent that increased storage costs are due to continued storage in spent fuel pools, indications are that switching to dry cask storage technologies will reduce costs. For example, the Sacramento Municipal Utility District (SMUD) estimates that the implementation of dry storage at its shut down Rancho Seco nuclear plant will reduce annual storage costs by almost 75 percent, from $10 million to $2.6 million.[21]

In sum, extended onsite storage would provide significant benefits in terms of reduced risks of environmental harm. It may also provide cost reductions in terms of overall repository costs, for given performance criteria, without increasing interim costs significantly.

Long-Lived "Low-Level" Waste Disposal

Of course, increasing repository capacity requirements to accommodate all long-lived wastes will add extra costs to the repository. As the analysis in Section B of this chapter "Component 1: Waste Reclassification" shows, disposing of long-lived low-level wastes from commercial power will add from 140 to 1,200 acres to repository area requirements. There would be additional costs from the additional nuclear weapons complex low-level wastes that would be reclassified as long-lived wastes, and from additional repository needs for transuranic wastes currently not designated for any repository.

Thus, it is clear that this restructuring will likely mean that net costs of long-lived nuclear waste management will increase over what is now anticipated. This should be balanced first of all against the greatly increased protection of the health and environment implied in our approach. The Department of Energy has a long history of taking the cheapest road in the short term only to come to huge costs in the long term both in financial and environmental terms. We should also note that the costs of managing the remaining, truly short-lived wastes may decrease somewhat since radioactivity would be far lower.

---

21. Personal communication from Rita Bowser, SMUD, Rancho Seco Nuclear Generating Station, (209) 333-2935 to Deborah Landau, IEER (21 February 1991).

Decommissioning Costs

The NRC, in developing decommssioning regulations, has provided estimates of what it will cost to decommission nuclear plants. These estimates are shown in Table 14, below, for immediate decommissioning, as well as for two deferred-dismantlement scenarios (after 30 years, and after 100 years).[22]

Table 14
NRC REACTOR DECOMMISSIONING COST ESTIMATES*

		SAFSTOR	
	DECON	**30 Years**	**100 Years**
PWR	$103 million	$101 million	$ 80 million
BWR	$132 million	$131 million	$106 million

Source: NRC 1988a, p. 15-5.
* Estimates are in 1986 dollars. Includes a 25 percent contingency factor; does not include cost for demolition of non-radioactive structures.

The total absolute costs depicted in this table are likely to be significant underestimates. According to a recent survey of the nuclear utility industry, for example, the total cost to decommission all 124 of the remaining U.S. reactors will be $25.6 billion (1990 dollars), or over $200 million per reactor.[23] The majority of utilities surveyed said they were planning on immediately decommissioning their reactors upon shut-down.

This is also readily apparent if we examine projected low-level waste disposal costs alone. For example, based on the DOE projection of about 18,000 cubic meters of waste generated by immediate decommissioning of a reference pressurized water reactor (PWR), and on

22. NRC 1988a, p. 15-5.
23. Borson 1990.

future low-level waste disposal costs ranging from $40 to $700 per cubic foot, low-level waste disposal costs alone from immediate decommissioning will range from $24 million to over $400 million.

In one respect, however, the numbers in the above table may be indicative: deferring decommissioning 50 to 100 years may greatly reduce the waste disposal cost component of decommissioning, due to the factor of 10 reduction in decommissioning waste volume. Thus, although the absolute costs are likely to be much higher than the NRC anticipates, the relative cost differences may still favor a 50- to 100-year delay in reactor decommissioning. These factors would tend to reduce certain portions of decommissioning costs, but there are other costs associated with delayed dismantlement, including monitoring and security measures, which tend to increase decommissioning costs.

A nuclear industry analysis of decommissioning costs indicates that on balance, delayed dismantlement after 100 years would cost slightly more than immediate dismantlement. The figures for a reference 1000-MW(e) pressurized water reactor are as follows:

Table 15
NUCLEAR INDUSTRY REACTOR DECOMMISSIONING COST ESTIMATES[*]

	Immediate	After 30 Years	After 100 years
Preparation	-	$ 30 million	$ 80 million
Surveillance	-	$ 35 million	$ 10 million
Removal	$180 million	$160 million	$120 million
TOTAL	$180 million	$225 million	$210 million

Source: CEA 1988, p.16.
* It appears that this estimate is in undiscounted constant dollars. Discounting costs to present value (a standard practice in economic analysis) would make the 100-year deferral the least expensive option, even if a low annual discount rate, such as 1 or 2 percent, was chosen.

Thus, based on NRC and nuclear industry estimates, it appears that delaying the decommissioning of commercial nuclear reactors will result in costs ranging from about 20 percent less (NRC) to about 15 percent more (industry) than immediate decommissioning.

We are inclined to think that the costs for delayed decommissioning are likely to be the same or less than immediate decommissioning. Recently, for example, the International Committee on Radiological Protection (ICRP) recommended that maximum occupational exposure limits be lowered from 5 rems per year to 2 rems per year.[24] Since delaying decommissioning is expected to result in a reduction of doses to workers involved in the process,[25] the relative costs of immediate dismantlement are likely to be far higher than now anticipated if the ICRP recommendations are adopted, as they typically are. The costs associated with worker protection, the risks of accident or significant environmental contamination, are all likely to be higher for immediate decommissioning than for deferred decommissioning.

In any case, the main point with regard to decommissioning is that there is no reason to believe there will be a significant increase in costs associated with the deferment of reactor decommissioning implied by the approach to waste management discussed here.

Disposal of Military Wastes

We do not have a basis at this time to estimate the impacts of our recommendations on the costs of waste management and disposal within the DOE weapons complex. However, we do make an observation regarding the lessons that the DOE's past waste management practices should hold for the future.

Since the widespread environmental contamination problems first began gathering increased public attention several years ago, the DOE has begun a "30-year" cleanup program for its weapons production complex, under the auspices of a new office, the Office of Environmental Restoration and Waste Management. The Department of Energy's current budget for waste management and environmental restoration is currently on the order of $5 billion to $6 billion per year.[26] Since there

---

24. ICRP 1991. The ICRP is an international scientific and policy committee which has been a principal force behind the development of radiation protection standards since its establishment in 1950.
25. NRC 1988a, p. 2-22.
26. From 1992 to 1997, the DOE's estimated budget for environmental restoration and waste management is slated to go from $4.4 to $6.7 billion per year. According to the DOE, actual annual needs might be between $8 and $9 billion per year from 1995 to 1997. DOE 1991b, p. 31.

are few new production activities going on, the bulk of this budget is for managing waste from past operations and site restoration activities, as well as wastes from the "cleanup" operation itself.

The overall budget for these activities over the next thirty years at the present rate would be about $150 billion. In essence this figure represents an order-of-magnitude estimate of the true cost of the DOE's past waste disposal practices, for it is largely these inadequate disposal practices of the past that have produced the problems of today. As DOE Secretary Admiral James Watkins said, "... the chickens have finally come home to roost, and years of inattention to changing standards and demands regarding the environment, safety and health are vividly exposed to public examination..."[27]

It is difficult to project the actual costs of the management of the various components of radioactive wastes which are currently present in the nuclear weapons complex. However, taken as a whole and considered as a waste disposal bill, the $150 billion is a huge sum, certainly far greater than the amount that would have been expended had adequate and careful attention been given to the disposal job from the beginning.

Thus, although we cannot make specific cost estimates for our approach in the DOE complex, we take a lesson from the past and observe that doing it right the first time is almost always inexpensive, in comparison to having to clean up the mess of a botched job after the fact.

Summary

The current program not only carries high risks of financial and programmatic failure, but even if it does manage to build a repository and emplace waste there, it still carries a significant risk of environmental harm that is far greater than a program based on a sound management approach.

Thus, the principal reason for adopting a new approach is to reduce the environmental risks associated with the current flawed program. However, our brief consideration also indicates that our approach is likely to be a prudent one from the point of long-term economic considerations as well.

---

27. Chronicle 1989, p. AT-114.

Although our proposed approach may cost more than *current projections* of total program costs, the history of DOE cost estimates indicates that the *actual* cost of the program if it continues to proceed along its present path will be far higher than the DOE currently anticipates. Indeed, planning *now* for a new long-lived waste management and disposal program with a much expanded time frame by implementing deliberately selected interim management alternatives *now* is likely to cost far less than if the current misguided program continues for 10 more years before collapsing in failure. At that point, the same alternatives will be forced upon the U.S. program at yet greater net cost, since billions of dollars spent in the interim will have been wasted. This is not mere hypothesis. The history of nuclear waste management is riddled with costly failures and remedial programs to correct these failures which are far more expensive than the initial outlays.

In the long run therefore, we believe our approach, if properly executed, will result in reduced environmental and public health risks at costs appropriate to the magnitude of the task at hand.

Chapter 5

SUMMARY AND RECOMMENDATIONS

Summary of Findings

1. Radioactive waste is inappropriately defined.

There is a fundamental problem with the way in which current government regulations categorize radioactive waste. For example, despite what is implied by their names, the two categories of waste named "high-level waste" and "low-level waste," are defined without systematic reference to their actual radioactivity levels. Instead, they are defined solely by the process which produced them. "High-level waste" is defined as spent reactor fuel, or those wastes resulting from the reprocessing of spent reactor fuel. "Low-level waste" is actually a catch-all category that is defined simply to include all radioactive waste that is not high-level waste, transuranic wastes, or uranium mill tailings.

Thus, the current radioactive waste categorization is in the untenable situation of sometimes labeling as "low-level" radioactive wastes which are actually several times more radioactive than other streams of radioactive waste which the current system labels "high-level." So, for example, the average radioactivity in the most radioactive portion of commercial low-level wastes (at 300 curies per cubic foot) is actually three times more radioactive than the average radioactivity in high-level wastes from nuclear weapons production activities. Even a typical reactor stream of low-level waste, which is routinely buried in shallow land trenches, is significantly more radioactive than some of these military high-level wastes. Many of the longer-lived and more dangerous categories of low-level waste, which is disposed of in shallow land burial, is also more radioactive than transuranic waste, which has been designated for disposal in a deep geologic repository.

A related problem with the existing categorization of radioactive waste is that it is without reference to the longevity of the waste's radioactivity. Both high-level and low-level wastes as currently defined can contain significant quantities of both long- and short-lived radionuclides. Since essential aspects of the radioactive waste disposal problem are in large part determined by the longevity of the waste's hazard, this also makes little sense.

2. Existing regulations and plans for long-lived radioactive waste management and disposal are irrational and incoherent.

Improper categorization of radioactive waste has been a principal obstacle to rational waste management policies. A central problem has been the substantial quantities of long-lived wastes involved in "low-level" waste categories. For example, long-lived plutonium-239 and other radionuclides have leaked from the now-closed low-level waste disposal facility at Maxey Flats, Kentucky.

Currently operative regulations for the disposal of low-level wastes, although much-improved since the time of Maxey Flats operation, continue to be fundamentally irrational. They are internally inconsistent, and therefore clearly inadequate. For example, the Nuclear Regulatory Commission (NRC) regulations currently require institutional controls at low-level waste disposal facilities for up to 100 years, because, according to NRC, "low-level" waste Classes A and B will decay to the point where they will present "an acceptable hazard" to any later intruder by the end of this time frame. However, this statement is logically and physically incompatible with the numerical limits also contained in those same NRC regulations. In fact, some forms of waste, if retrieved from the disposal site after the 100-year period had elapsed, and then re-buried as if for the first time, would have levels, according to the same NRC regulations, such as to require a 100-year institutional control period all over again. Indeed, even according to the NRC regulation's own definitions of what is "hazardous" and what is "acceptable," wastes could be buried which will be unacceptably hazardous for thousands of years beyond the time when the regulations say they should pose an "acceptable hazard." Hence, the internal inconsistency of the regulations and definitions.

The U.S. Environmental Protection Agency (EPA) has the authority under law to promulgate low-level waste standards, and has actually formulated comprehensive standards for this purpose, but disagreements with the NRC and the Department of Energy (DOE) arising due to the fact that the EPA's standards are more comprehensive and stringent have prevented their official publication. EPA's low-level waste standards are intended to provide comprehensive and consistent coverage across both commercial and military facilities, and to protect groundwater. Since existing standards are severely lacking in these areas, the lack of EPA low-level waste standards is a serious defect in the U.S. low-level waste program.

While radioactive waste management is a difficult issue by its very nature, it does not have to be irrational. For instance, radioactive waste management in Sweden is based on the principle that radioactive waste disposal methods should be determined by the longevity of the waste. Thus, long-lived wastes (whether they would be considered "low-level" or "high-level" in the U.S.) are slated for disposal in a deep geological repository. As a consequence, fully 40 percent of the volume slated for disposal in Sweden's projected long-lived waste repository consists of reactor wastes which would be considered "low-level" in the U.S.

3. The Department of Energy's management of the repository program for long-lived radioactive wastes is exacerbating these problems.

The U.S. Department of Energy is responsible for developing geologic repositories for high-level and transuranic wastes. Over the past 15 years, timetables for both of the DOE's major repository programs have slipped and costs have escalated. For example, an operational repository for high-level wastes is now, according to DOE projections, twice as far away as it was in the late 1970's, and projected disposal costs grew by over 80 percent between 1983 and 1990.

Despite the existence of at-reactor and onsite storage options for nuclear wastes and repeated failure within its repository program, a sense of urgency continues to pervade the U.S. attitude towards long-lived radioactive waste disposal. The nuclear industry in particular is anxious to have in place a program which will allow the government to take the waste out of their hands. As one nuclear utility executive recently put it, the government should take charge of spent fuel waste by 1998 (a target date in the 1982 Nuclear Waste Policy Act), and, he said, "I don't care where you put it."

This situation is exacerbated by lack of NRC and EPA standards for repository disposal which correspond to longevity of the wastes and the health threats posed by many long-lived radionuclides.

4. Taken as a whole, current policies entail high risks in terms of both economics and environmental protection.

Economics

On the high-level waste side, where $3 billion has already been spent on a program for geologic repository disposal, projections of disposal cost per ton have increased by over 80 percent since 1983 (in constant dollars), from $179,100 per metric ton, to about $325,000 per metric ton.

The cost of the WIPP repository program, which should be more stable since the program is further along,shows similar increases. In just two years, the DOE's cost estimates for the five-year period including the first several years of WIPP operation have more than doubled, from $531 million in 1989, to around $1.1 billion in 1991.

Seven hundred million dollars have already been spent on development of the low-level waste disposal sites according to fundamentally inadequate environmental standards. Most of these sites will likely miss legal deadlines established for their availability.

Additional billions are being spent and will have to be spent to fix the problems from past shallow-land burial of low-level wastes at both commercial and military sites.

For instance, at the contaminated commercial low-level waste disposal facility at Maxey Flats, Kentucky, when the clean-up is finally done, and all the costs accounted for, final disposal costs for wastes disposed of there may well be 10 to 50 times greater than the original disposal rates. At West Valley, New York, the bill for combined high-level waste management and remediation of problems with low-level waste disposal is now estimated to cost between $2.4 and $3.4 billion.[1] These wastes resulted from a reprocessing plant for plutonium extraction which originally cost $35 million and was supposed to be a commercial, profit-making operation.[2] Finally, a large proportion of the

---

1. Cost estimates provided to Senator Daniel Patrick Moynihan (D-NY) by aides to Secretary of Energy James Watkins, as reported in Douglas Turner, "Cost of West Valley Nuclear Clean-up Soars," Buffalo News, Section B, p. 1 (February 21, 1991). Reference courtesy of Carol Mongerson, Coalition on West Valley Nuclear Wastes, East Concord, New York.
2. Carter 1987, p. 98.

$150 billion cost of clean-up of the nuclear weapons complex sites is due to environmental threats created by dumping of radioactive and mixed wastes into the soil and the high-level wastes now sitting in 228 large tanks at Hanford and the Savannah River Site.

Billions of dollars have been or are slated to be spent stabilizing uranium mill tailings and preventing radium-226 and thorium-230 from getting into the groundwater. Yet, such programs are limited to 1,000 years of environmental protection, despite the fact that hazards will persist for the several hundred thousand years which will be necessary for the thorium-230 to decay substantially.

Environmental Protection

On the high-level waste side, the DOE's program has become environmentally more risky due to the reduction in the number of potential repository sites to be characterized to one, despite its greater cost. The selection of Nevada as the only site to be characterized came about as a result of a process that, in part, started with the fact that the government already controlled the land, and ended in a decision in which politics overwhelmed science.

The potential doses from vitrified military high-level wastes alone, a small portion of the total radioactivity proposed to be disposed of, could far exceed allowable standards due to the potential incompatibility of glass with the Yucca Mountain hydrogeological conditions.

The huge inventories of transuranic wastes in the form of contaminated soil which will be left unaddressed by the WIPP program means that this program is guaranteed to be a failure from the point of view of isolating transuranic wastes from the human environment. Moreover, the small proportion that may be disposed of in WIPP is by no means assured of isolation from the human environment even for a few thousand years, much less the hundreds of thousands of years it will remain threatening.

Likewise, despite huge cost escalations in the development of low-level waste disposal sites (unit disposal costs for some disposal sites in the 1990s are projected to be 600-700 times 1975 disposal costs), these still entail huge environmental risks for future generations because of the fundamental inadequacy and irrationality of the underlying regulations.

In sum, for a host of fundamental reasons, it is highly unlikely that minimization of risk to future generations or wise use of financial resources can be achieved under the present approaches for management of any category of nuclear waste. Far more likely is waste of money coupled with a festering problem, and potentially increasing risks from inappropriate policies.

Recommendations

The management of long-lived, highly radioactive waste has no safe or simple solution. It is a difficult, messy and costly problem. We have created a problem which will be a source of substantial risk to future generations, whatever we do today. It is therefore imperative that society minimize the generation of further long-lived radioactive wastes. Almost all long-lived wastes, in terms of the quantity of radioactivity involved as well as the physical volume, come from nuclear power and nuclear weapons production. The problem of the management of long-lived radioactive wastes makes it incumbent upon us that any further generation of long-lived wastes from these two sources, about which there is no social consensus, be subjected to careful scrutiny regarding the potential for phase-out. Generation of long-lived wastes from medical and research facilities should also be minimized by use of shorter-lived isotopes and substitute processes to the extent practicable.

It is of vital importance that we address the problem of the wastes which already exist with the greatest scientific and technical integrity so that future generations may be protected to the greatest extent possible. Steps must also be taken in the interim to see that the present generation does not suffer large releases and contamination due to accidents or poor interim management.

Based on the above findings and these general considerations, we make three overall recommendations regarding the U.S. radioactive waste management system: change radioactive waste classifications so that all long-lived wastes are managed according to an integrated hazard-and-longevity-based approach, restructure the program for disposing of these newly defined long-lived wastes, and provide for extended onsite storage of wastes in the interim while the long-term problem is being addressed. Each of these entails a number of sub-recommendations, which are discussed below along with some of their ramifications.

1. Change how radioactive wastes are defined, and reclassify radioactive wastes and their disposal according to longevity and hazard level.

Since many of the problems of the current waste management system are due to the underlying fundamental inadequacy of waste definitions, an integrated approach which entails redefining wastes according to their longevity is needed. This includes reclassification of considerable quantities of commercial and military wastes that are now considered "low-level" into the long-lived category. Long-lived should be appropriately defined such that wastes containing significant quantities of cesium-137 and strontium-90 are considered long-lived. Evidently, what constitutes "significant" must be determined by health-based criteria and this must be the subject of careful scientific study and broad democratic debate.

An important consequence of this is to expand the amount of waste being sent to a repository. By analogy to the Swedish approach, we expect that approximately 225,000 cubic meters of commercial reactor waste now considered low-level would go to a deep repository. Depending on the emplacement density of this waste, we expect that this amount would require an additional 140 to 1,200 acres of repository space in addition to the approximately 2,400 acres already needed for spent fuel and reprocessing wastes. Although this does not include the consequences from reclassifying military low-level wastes, it is indicative of the scale of the problem.

The DOE made a partial step in this direction when it decided in 1970 to reclassify some of its "low-level" wastes as transuranic wastes and dispose of them in a repository. However, the application of this principle has been seriously deficient. First, the repository selection, characterization, and testing has been flawed. Second, the volume of wastes which fall into this category is far larger than the proposed repository in New Mexico will be able to accommodate.

2. Restructure the entire long-lived waste management and disposal program.

The present process of site selection and characterization for the high level and transuranic waste repositories has been compromised both technically and institutionally. It has become necessary to abandon it. The low-level waste disposal site selection programs, are a welter of confusion based on an irrational NRC waste classification scheme that mixes intensely radioactive long-lived waste with short-lived waste, and the absence of any applicable EPA standards at all. There is no provision for long-term isolation of the long-lived components of uranium mill tailings.

The present programs for selection of disposal sites must be abandoned and replaced with an approach to long-lived waste management and disposal that has technical integrity and institutional competence.

Regarding the programs established for each of the categories of radioactive waste, our suggested alternative approach entails:

- *Spent Fuel and High-Level Reprocessing Wastes:* Cancellation of the high-level waste repository development program as currently being implemented by the DOE. All current investigations of Yucca Mountain should be halted, and the program should begin again with basic consideration of geology, and rock types, as well as consideration of alternative approaches such as sub-seabed disposal and, for already reprocessed wastes, transmutation.

- *Transuranic Wastes:* Along lines similar to those which apply to the Yucca Mountain program, a cancellation of the transuranic waste repository program at WIPP.

- *Low-Level Wastes:* Cancellation of siting for new low-level waste sites, with provision for monitored storage of low-level wastes, study of the feasibility of separating all long-lived components from them, as well as consideration of the feasibility of storing hospital and pharmaceutical and research wastes at the most appropriate, interim locations.

  At a minimum, no siting, construction or operation of new low-level waste disposal facilities should be allowed to proceed in the absence of comprehensive EPA standards for low-level waste disposal.

- *Uranium Mill Tailings:* An assessment of the feasibility of separating radium-226 and thorium-230 from uranium mill tailings to enable their integration into the long-term management program.

Addressing related institutional and policy issues entails a number of other steps, including:

- The removal of the DOE from the waste management program and the establishment of an independent radioactive waste management authority which does not have conflicts between nuclear power and weapons production on the one hand, and environmental and health protection on the other.

(This includes the creation of new institutional arrangements for managing the research and development needed to create a program for long-term management of long-lived wastes.)

- Implementation of policies designed to create substantial incentives for actions to minimize the generation of long-lived radioactive wastes.

- The establishment of consistent, health-based standards governing nuclear waste management and disposal for all radioactive waste, irrespective of the process producing those wastes.

3. Provide for extended onsite storage of long-lived radioactive wastes at the point of generation as an interim management step.

In order to accommodate the needs of a restructured development program for long-term waste management, extended onsite storage will be needed for various categories of waste. This would include:

- Planning to allow up to 100 years of at-reactor storage of spent fuel (in dry casks) and other long-lived radioactive wastes to accommodate the reality that a long-term waste isolation option will not be available for many decades. Funds for extended onsite storage should come from the Nuclear Waste Fund.

- Likewise, planning to defer decommissioning of shut-down nuclear reactors by up to 100 years to lower disposal requirements, reduce risk, and integrate onsite storage with a realistic time frame of radioactive waste disposal.

- The stabilization of radioactive wastes including military high-level, long-lived low-level and transuranic wastes for storage on site in a manner that reduces the risk to workers and residents and which does not compromise in any essential way long-term management programs which may be put into place. Retrievable and carefully monitored storage in solid form in ways which minimize the risk of contamination of soil, water and air should guide the examination of options for interim storage.

Combined with onsite storage and deferral of decommissioning for nuclear reactors, a restructured long-lived waste disposal program will allow time for development of a careful and sound understanding of geology and climate factors affecting disposal options, development of waste forms with better isolation characteristics and research on new technologies. This will allow the science to be done in parallel with the politics, in contrast to the present program where politics and policy goals in areas other than public health and environmental protection have tended to dominate.

BIBLIOGRAPHY AND REFERENCES

ADL, 1990
Arthur D. Little, Inc., *Exposure Assessment of Carbon-14 Releases from Disposal of Spent Nuclear Fuel in an Underground Repository in Tuff*, Submitted to U.S. Environmental Protection Agency, September.

Berlin, 1989
Robert E. Berlin and Catherine Stanton, *Radioactive Waste Management*, New York: John Wiley and Sons, Inc.

Borson, 1990
Daniel Borson, et al., *Payment Due: A Reactor-by-Reactor Assessment of the Nuclear Industry's $25+ Billion Decommissioning Bill*, Washington, DC: Public Citizen, October 11.

Carter, 1987
Luther J. Carter, *Nuclear Imperatives and Public Trust: Dealing with Radioactive Waste*, Washington, DC: Resources for the Future.

CEA, 1988
U.S. Council for Energy Awareness, *Completing the Task, Decommissioning Nuclear Power Plants*, Washington, DC: CEA, October.

Chem-Nuclear, 1991
Chem-Nuclear Systems, Inc., "Barnwell Low-Level Radioactive Waste Management Facility Rate Schedule," memorandum, effective January 1.

Chronicle 1989
San Francisco Chronicle, June 28.

Davis, 1991   Dave Davis, "Ohio Assigned To Build Region's Nuclear Dump," *Cleveland Plain Dealer*, July 25.

DOE, 1991a   U.S. Department of Energy, *Draft Mission Plan Amendment*, DOE/RW-0316P, September.

DOE, 1991b   U.S. Department of Energy, *Environmental Restoration and Waste Management Five-Year Plan, Fiscal Years 1993-1997*, DOE/S-0090P, Washington, DC: DOE, August.

DOE, 1990a   U.S. Department of Energy, *Preliminary Estimates of the Total-System Cost for the Restructured Program: An addendum to the May 1989 Analysis of the Total-System Life Cycle Cost for the Civilian Radioactive Waste Management Program*, DOE/RW-0295P, Washington, DC: DOE Office of Civilian Radioactive Waste Management, December.

DOE, 1990b   U.S.Department of Energy, *1989 State-by-State Assessment of Low-Level Radioactive Wastes Received at Commercial Disposal Sites*, DOE/LLW-107, Idaho Falls, ID: National Low-Level Waste Management Program, December.

DOE, 1990c   U.S. Department of Energy, *1989 Annual Report on Low-Level Radioactive Waste Management Progress*, DOE/EM-0006P, Washington, DC: DOE Office of Environmental Restoration and Waste Management, October.

DOE, 1990d   U.S. Department of Energy, *Integrated Data Base for 1990: U.S. Spent Fuel and Radioactive Waste Inventories, Projections, and Charateristics*, DOE/RW-0006, Rev. 6, prepared by Oak Ridge National Laboratory for the Office of Civilian Radioactive Waste Management and the Office of Environmental Restoration and Waste Management, Washington, DC: U.S.DOE, October.

DOE, 1990e   U.S. Department of Energy, Office of Civilian Radioactive Waste Management, Presentation entitled "Requirements for Controlling a Repository's Releases of Carbon-14 Dioxide; The High Cost and Negligible Benefits," October 26.

DOE, 1990f   U.S. Department of Energy, *Final Supplemental Environmental Impact Statement*, Waste Isolation Pilot Plant, January.

DOE, 1990g   U.S. Department of Energy, *Final Safety Analysis Report, Waste Isolation Pilot Plant*, January.

DOE, 1989a   *U.S. Department of Energy, Environmental Restoration and Waste Management Five-Year Plan*, DOE/S-0070, Washington, DC: U.S.DOE.

DOE, 1989b   U.S. Department of Energy, *Analysis of the Total System Life Cycle Cost for the Civilian Radioactive Waste Management Program*, DOE/RW-0236, Washington, DC: DOE Office of Civilian Radioactive Waste Management, May.

DOE, 1989c   U.S. Department of Energy, *Draft Supplemental Environmental Impact Statement*, Waste Isolation Pilot Plant, DOE/EIS-0026-DS, April.

DOE, 1989d   U.S. Department of Energy, *Final Version Dry Cask Storage Study*, DOE/RW-0220, Washington, DC: DOE Office of Civilian Radioactive Waste Management, February.

DOE, 1988a   U.S. Department of Energy, *Site Characterization Plan Overview*, DOE/RW-0198, Washington, DC: DOE Office of Civilian Radioactive Waste Management, December.

DOE, 1988b   U.S. Department of Energy, *Radioactive Waste Management*, DOE Order 5820.2A, September.

DOE, 1988c   U.S. Department of Energy, *Integrated Data Base for 1988: U.S. Spent Fuel and Radioactive Waste Inventories, Projections, and Characteristics*, DOE/RW-0006, Rev. 4, prepared by Oak Ridge National Laboratory, Washington, DC: U.S. DOE, September.

DOE, 1987   U.S. Department of Energy, 10 CFR Part 962, *Federal Register*, Vol. 52, Washington, DC: GPO, May 1.

DOE, 1984   U.S. Department of Energy, *Spent Fuel and Radioactive Waste Inventories, Projections, and Characteristics*, DOE/RW-006, prepared by Oak Ridge National Laboratory, Washington, DC: U.S. DOE, September.

DOE, 1979   U.S. Department of Energy, *Management of Commercially Generated Radioactive Waste*, DOE/EIS-0046-D, Washington, DC: U.S. DOE, April.

Eisenbud, 1987   Merril Eisenbud, *Environmental Radioactivity from Natural, Industrial and Military Sources*, Third Edition, San Diego, CA: Academic Press.

EPA, 1991   U.S. Environmental Protection Agency, "Environmental Standards for the Management and Disposal of Spent Nuclear Fuel, High-Level and Transuranic Radioactive Wastes," 40 CFR Part 191, draft proposed rule, April 26.

EPA, 1990a   U.S. Environmental Protection Agency, "The Final Conditional No-migration Determination for the Department of Energy's Waste Isolation Pilot Plant," *Federal Register*, Vol. 55, November 14.

EPA, 1990b   Caroline Petti and James Gruhlke, U.S. Environmental Protection Agency, "EPA's Draft Environmental Standard for Low-Level Radioactive Waste Management and Disposal," for presentation at the New England Environmental Conference, March.

EPA 1990c   U.S. Environmental Protection Agency, Code of Federal Regulations, 40 CFR Parts 190-299, Washington, DC: GPO, January.

EPA, 1989   U.S. Environmental Protection Agency, "Repromulgation of U.S. EPA's Environmental Standards for Disposal of Spent Nuclear Fuel, High-Level, and Transuranic Radioactive Wastes," by Daniel Egan, Raymond Clark, Floyd Galpin, and William Holcomb, for presentation at the Waste Management '89 Symposium, February 26 to March 3.

EPA, 1986   U.S. Environmental Protection Agency, *RCRA Orientation Manual*, Washington, DC: U.S. EPA, January.

EPA, 1985   U.S. Environmental Protection Agency, "Environmental Radiation Protection Standards for Management and Disposal of Spent Nuclear Fuel, High-Level and Transuranic Radioactive Wastes; Final Rule" 40 CFR Part 191, *Federal Register*, Vol. 50, No. 1982, Washington, DC: GPO, September 19.

EPA, 1983a   U.S. Environmental Protection Agency, "Environmental Standards for Uranium and Thorium Mill Tailings at Licensed Commercial Processing Sites: Final Rule," 40 CFR 192, Subpart D, *Federal Register*, Vol. 48, No. 196, Washington, DC: GPO, October 7.

EPA, 1983b   U.S. Environmental Protection Agency, *Final Environmental Impact Statement for Standards for the Control of Byproduct Materials from Uranium Ore Processing*, 40 CFR Part 192, Vol. 1, EPA 520/1-83-008-1, Washington, DC: EPA Office of Radiation Programs, September.

EPA, 1983c   U.S. Environmental Protection Agency, "Environmental Standards for the Control of Residual Radioactive Materials from Inactive Uranium Processing Sites: Final Rule," 40 CFR Part 192, Subpart A, *Federal Register*, Vol. 38, Washington, DC: GPO, January 5.

GAO, 1991   U.S. General Accounting Office, *Delays in Addressing Environmental Requirements and New Safety Concerns Affect DOE's Waste Isolation Pilot Plant*, GAO/T-RCED-91-67, Testimony of Victor S. Rezendes, Director of Energy Issues, Resources, Community, and Economic Development Division, before the Environment, Energy and Natural Resources Subcommittee, Committee on Government Operations, U.S. House of Representatives, Washington, DC, June 13.

GAO, 1990   U.S. General Accounting Office, *Changes Needed in DOE User-Fee Assessments to Avoid Funding Shortfall*, GAO/RCED-90-65, June.

GAO, 1988a   U.S. General Accounting Office, *Fourth Annual Report on DOE's Nuclear Waste Program*, GAO/RCED-88-131.

GAO, 1988b   U.S. General Accounting Office, testimony by Keith O. Fultz, Senior Associate Director, Resource, Community and Economic Development Division, *Status of the Department of Energy's Waste Isolation Pilot Plant*, GAO/T-RCED-88-63, September 13.

GAO, 1988c   U.S. General Accounting Office, *Problems Associated with DOE's Inactive Waste Site*, GAO/RCED-88-169, August.

GAO, 1986a   U.S. General Accounting Office, *Environmental Issues at DOE's Nuclear Defense Facilities*, GAO/RCED-86-192, September.

GAO, 1986b   U.S. General Accounting Office, *Monitored Retrievable Storage of Spent Nuclear Fuel*, GAO/RCED-86-104FS, May.

GAO, 1986c   U.S. General Accounting Office, *Nuclear Waste: Department of Energy's Transuranic Waste Disposal Plant Needs Revision*, GAO/RCED-86-90, March.

Goessl, 1990	Joan Goessl, "Slab Fall Stirs Safety Review at WIPP Site," the Associated Press, June 12.
Harvey, 1991	Jennifer Harvey, et al., "Status of Spent Nuclear Fuel Storage," National Association of Regulatory Utility Commissioners, July 20.
Hollister, 1981	Charles Hollister, D. Richard Anderson, and G. Roth Heath, "Subseabed Disposal of Nuclear Waste," *Science*, Vol. 213, No. 4514, September 18, 1981.
House 1991	U.S. House of Representatives, *Proposals Relating to the Operation of the Waste Isolation Pilot Plant in New Mexico*, Hearing before the Energy and Environment Subcommittee of the Committee on Interior and Insular Affairs, April 16.
House, 1988	U.S. House of Representatives, *Status of the Waste Isolation Pilot Plant Project*, Hearing before a Subcommittee on Government Operations, September 13.
ICRP, 1991	International Commission on Radiological Protection, "1990 Recommendations of the International Commission on Radiological Protection," *Annals of the ICRP*, Publication 60, Vol. 27, No. 1-3, New York: Pergamon Press.
Jordan, 1984	Julie M. Jordan, National Conference of State Legislatures, *Low-Level Radioactive Waste Management: An Update*, October.
Lipshutz, 1980	Ronnie D. Lipshutz, *Radioactive Waste: Politics, Technology and Risk,* A Report of the Union of Concerned Scientists, Cambridge, MA: Ballinger Publishing Co.
LLRWPA, 1980	The Low-Level Radioactive Waste Policy Act of 1980, Public Law 96-57, December 22.
LLRWPAA, 1986	The Low-Level Radioactive Waste Policy Amendments Act of 1985, Public Law 99-240, January.

Makhijani, 1991   Arjun Makhijani, *Glass in the Rocks: Some Issues Concerning the Disposal of Radioactive Borosilicate Glass in a Yucca Mountain Repository*, Takoma Park, MD: Institute for Energy and Environmental Research, January 29.

Makhijani, 1989   Arjun Makhijani, *Reducing the Risks: Policies for the Management of Highly Radioactive Nuclear Waste*, Takoma Park, MD: Institute for Energy and Environmental Research, May.

Mansure, 1985   A.J. Mansure, "Underground Facility Area Requirements for a Radioactive Waste Repository at Yucca Mountain," SAND84-1153, Albuquerque, NM: Sandia National Laboratory, November.

MRSRC, 1989   Monitored Retrievable Storage Review Commission, *Nuclear Waste: Is There a Need for Federal Interim Storage?*, Washington, DC: GPO, November.

NAS, 1990   National Academy of Sciences National Research Council, *Rethinking High-Level Radioactive Waste Disposal*, Washington, DC: National Academy Press.

NAS, 1986   National Academy of Sciences National Research Council, *Scientific Basis for Risk Assessment and Management of Uranium Mill Tailings*, Washington, DC: National Academy Press.

NAS, 1984   National Academy of Sciences, *Review of the Swedish KBS-3 Plan for Final Storage of Spent Nuclear Fuel*, Washington, DC: National Academy Press, March.

NAS, 1983   National Academy of Sciences National Research Council, *A Study of the Isolation System for Geologic Disposal of Radioactive Waste*, Waste Isolation Systems Panel, Washington, DC: National Academy Press, 1983.

NAS, 1957   National Academy of Sciences, *The Disposal of Radioactive Waste on Land*, report of the Committee on Waste Disposal of the Division of Earth Sciences, September.

Nevada, 1989   Nevada Agency for Nuclear Projects, Nuclear Waste Project Office, *State of Nevada Comments on the U.S. Department of Energy Site Characterization Plan, Yucca Mountain Site, Nevada*, September.

NRC, 1991   U.S. Nuclear Regulatory Commission, "Standards for Protection Against Radiation: Final Rule," 10 CFR Part 20 et al., *Federal Register*, Vol. 56, No. 98, May 21.

NRC, 1990   U.S. Nuclear Regulatory Commission, "Storage of Spent Fuel in NRC-Approved Storage Casks at Power Reactor Sites," *Federal Register*, Vol. 55, No. 136, July 18.

NRC, 1989a   U.S. Nuclear Regulatory Commission, "Waste Confidence Decision Review," 10 CFR Part 51, *Federal Register*, Vol. 54, September 28.

NRC, 1989b   U.S. Nuclear Regulatory Commission, "Disposal of Radioactive Wastes," *Federal Register*, Vol. 54, No. 100, May 25.

NRC, 1988a   U.S. Nuclear Regulatory Commission, *Final Generic Environmental Impact Statement on Decommissioning of Nuclear Facilities*, NUREG-0586, Washington, DC: NRC Office of Nuclear Regulatory Research, August.

NRC, 1988b   U.S. Nuclear Regulatory Commission, 10 CFR, Parts 51-199, Washington, DC: GPO, January.

NRC, 1983a   U.S. Nuclear Regulatory Commission, "Licensing Requirements for the Land Disposal of Radioactive Waste," 10 CFR Part 61, Washington, DC: GPO.

NRC, 1983b   U.S. Nuclear Regulatory Commission, *Workshop on
                    Spent Fuel/Cladding Reaction During Dry Storage*,
                    NUREG/CP-0049, Washington, DC: NRC Office of
                    Nuclear Regulatory Research, August.

NRDC, 1987   Thomas B. Cochran, William M. Arkin, Robert S.
                    Norris, and Milton M. Hoenig (Natural Resources
                    Defense Council), *Nuclear Weapons Databook,
                    Volume II: U.S. Nuclear Warhead Production*, New
                    York: Ballinger Publishing Company.

NYSERDA,   New York State Energy Research and Development
1990   Authority, *New York State Low-Level Radioactive
                    Waste Status Report for 1989*, June.

Nucleonics,   "Giant Sponges and Sea-Dumped Radwaste: EPA
1977   Investigates," *Nucleonics Week*, December 27.

Nucleonics,   "Michigan Governor Reverses Pull-Out from LLW
1989a   Compact," *Nucleonics Week*, March 9.

Nucleonics,   "LLW Disposal Figures for 1988 Show 22.5%
1989b   Decline from 1987," *Nucleonics Week*, February 16.

OTA, 1991   U.S. Congress, Office of Technology Assessment,
                    *Long-Lived Legacy: Managing High-Level and
                    Transuranic Waste at the DOE Nuclear Weapons
                    Complex*, OTA-BP-O-83, Washington, DC: GPO,
                    May.

OTA, 1989   U.S. Congress, Office of Technology Assessment,
                    *Partnerships Under Pressure: Managing
                    Commercial Low-Level Radioactive Waste*, OTA-O-
                    426, Washington, DC: GPO, November.

OTA, 1988   U.S. Congress, Office of Technology Assessment,
                    *An Evaluation of Options for the Managing of
                    Greater-than Class-C Low-Level Radioactive
                    Waste*, OTA-BP-0-50, Washington, DC: GPO,
                    October.

OTA, 1986   U.S. Congress, Office of Technology Assessment, *Staff Paper on the Subseabed Disposal of High-Level Radioactive Waste,* Washington, DC: GPO, May.

Resnikoff, 1987   Marvin Resnikoff, *Living Without Landfills,* New York: Radioactive Waste Campaign.

Saleska, 1989   Scott Saleska, et al., *Nuclear Legacy: An Overview of the Places, Problems, and Politics of Radioactive Waste in the United States,* Washington, DC: Public Citizen, September.

Sandia, 1990   Sandia National Laboratories, *Preliminary Comparison with 40 CFR Part 191, Subpart B for the Waste Isolation Pilot Plant,* SAND90-2347, Springfield, VA: NTIS, December.

Schneider, 1989   Keith Schneider, "Nuclear Waste Dump Faces Another Potential Problem," *New York Times,* June 3.

SKB, 1990   Swedish Nuclear Fuel and Waste Management Company, *Plan 90: Costs for Management of the Radioactive Waste from Nuclear Power Production,* SKB Technical Report 90-33, June.

SRIC, 1985   Southwest Research and Information Center, "The Costs of Uranium: Whose Paying with Lives, Lands, and Dollars," *The Workbook,* Vol. 10, No. 3, July/September.

Szymanski, 1987   Jerry S. Szymanski, *Conceptual Considerations of the Death Valley Groundwater System with Special Emphasis on the Adequacy of This System to Accommodate the High-Level Nuclear Waste Repository,* Las Vegas: DOE Nevada Operations Office, November.

Trapp   John Trapp, U.S. Nuclear Regulatory Commission, "Probability of Volcanism at Yucca Mountain," memorandum, undated, but apparently late 1988 or early 1989.

US Ecology,   US Ecology, "Nevada Nuclear Center Schedule of
1990   Charges, Radioactive Waste," memorandum
                    effective August 1.

Van   R.A. Van Konynenburg, Lawrence Livermore
Konynenburg,   National Laboratory, "Gaseous Release of Carbon-
1991   14: Why the High Level Waste Regulations Should
                    Be Changed," *High Level Radioactive Waste
                    Management*, Proceedings of the Second Annual
                    International Conference, Las Vegas, Nevada, April
                    28-May 3, 1991, La Grange Park, IL: American
                    Nuclear Society, Inc.

Wald, 1989   Matthew Wald, "Work Is Faltering on U.S.
                    Repository for Atomic Wastes," *New York Times*,
                    January 17.

Warren, 1989   Jennifer Warren, "Monument to Nuclear Age:
                    Dump for Nuclear Waste," *Los Angeles Times*,
                    March 19.

Wilson, 1979   C.L. Wilson, "Nuclear Energy: What Went
                    Wrong?" *Bulletin of the Atomic Scientists*, June.

Wodrich, 1989   D.D. Wodrich, "NRC Concerns About Grouting
                    Double Shell Tank Waste," Westinghouse Hanford
                    Company memorandum 13300-89-DDW-039,
                    March 16.

About the Authors

Arjun Makhijani, the President of IEER, has a PhD in Engineering from the University of California at Berkeley. He is the author or coauthor of numerous studies on nuclear waste and other environmental issues.

Scott Saleska, a staff scientist at IEER, has a degree in Physics from MIT. He is the author of *Nuclear Legacy*, a study of radioactive waste in the United States.